my **revision** notes

AQA A

CHE STRY

Rob King

Editor: Graham Curtis

HODDER EDUCATION

Hodder Education, an Hachette UK company, 338 Euston Road, London NW1 3BH

Orders
Bookpoint Ltd, 130 Milton Park, Abingdon, Oxfordshire OX14 4SB
tel: 01235 827827
fax: 01235 400401
e-mail: education@bookpoint.co.uk
Lines are open 9.00 a.m.–5.00 p.m., Monday to Saturday, with a 24-hour message answering service. You can also order through the Hodder Education website: www.hoddereducation.co.uk

ISBN 978-1-4441-8080-0

First printed 2013
Impression number 5 4 3 2 1
Year 2017 2016 2015 2014 2013

Cover photo reproduced by permission of Gunnar Assmy/Fotolia

Typeset by Datapage (India) Pvt. Ltd.
Printed in India

P2188

Get the most from this book

Everyone has to decide his or her own revision strategy, but it is essential to review your work, learn it and test your understanding. These Revision Notes will help you to do that in a planned way, topic by topic. Use this book as the cornerstone of your revision and don't hesitate to write in it — personalise your notes and check your progress by ticking off each section as you revise.

☑ Tick to track your progress

Use the revision planner on pages 4 and 5 to plan your revision, topic by topic. Tick each box when you have:

- revised and understood a topic
- tested yourself
- practised the exam questions and gone online to check your answers and complete the quick quizzes

You can also keep track of your revision by ticking off each topic heading in the book. You may find it helpful to add your own notes as you work through each topic.

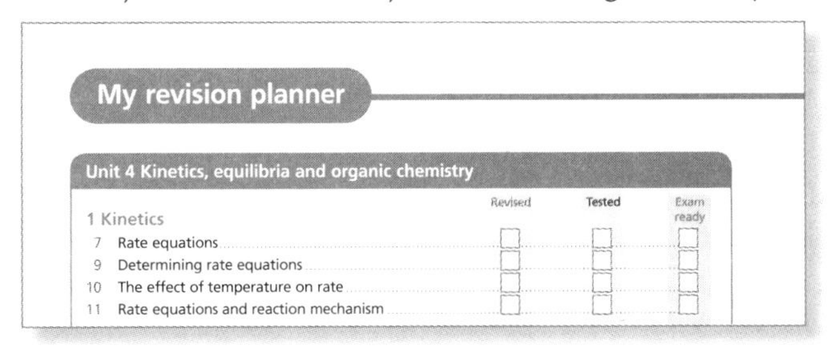
My revision planner

Unit 4 Kinetics, equilibria and organic chemistry

		Revised	Tested	Exam ready
1 Kinetics				
7	Rate equations	☐	☐	☐
9	Determining rate equations	☐	☐	☐
10	The effect of temperature on rate	☐	☐	☐
11	Rate equations and reaction mechanism	☐	☐	☐

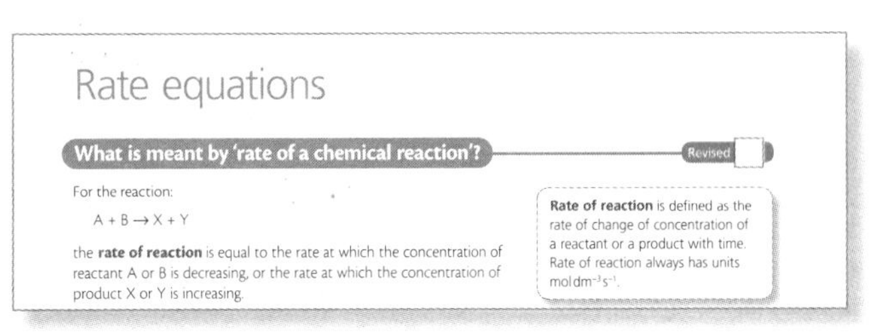
Rate equations

What is meant by 'rate of a chemical reaction'? Revised

For the reaction:

$A + B \rightarrow X + Y$

the **rate of reaction** is equal to the rate at which the concentration of reactant A or B is decreasing, or the rate at which the concentration of product X or Y is increasing.

Rate of reaction is defined as the rate of change of concentration of a reactant or a product with time. Rate of reaction always has units $mol\,dm^{-3}\,s^{-1}$.

Features to help you succeed

Examiners' tips and summaries

Expert tips are given throughout the book to help you polish your exam technique in order to maximise your chances in the exam.

The summaries provide a quick-check bullet list for each topic.

Typical mistakes

The author identifies the typical mistakes candidates make and explains how you can avoid them.

Now test yourself

These short, knowledge-based questions provide the first step in testing your learning. Answers are at the back of the book.

Definitions and key words

Clear, concise definitions of essential key terms are provided where they first appear.

Key words from the specification are highlighted in bold throughout the book.

Revision activities

These activities will help you to understand each topic in an interactive way.

Exam practice

Practice exam questions are provided for each topic. Use them to consolidate your revision and practise your exam skills.

Online

Go online to check your answers to the exam questions and try out the extra quick quizzes at **www.therevisionbutton.co.uk/myrevisionnotes**

My revision planner

Unit 4 Kinetics, equilibria and organic chemistry

Revised | Tested | Exam ready

Exam practice answers and quick quizzes at www.therevisionbutton.co.uk/myrevisionnotes

Countdown to my exams

6–8 weeks to go

- Start by looking at the specification — make sure you know exactly what material you need to revise and the style of the examination. Use the revision planner on pages 4 and 5 to familiarise yourself with the topics.
- Organise your notes, making sure you have covered everything on the specification. The revision planner will help you to group your notes into topics.
- Work out a realistic revision plan that will allow you time for relaxation. Set aside days and times for all the subjects that you need to study, and stick to your timetable.
- Set yourself sensible targets. Break your revision down into focused sessions of around 40 minutes, divided by breaks. These Revision Notes organise the basic facts into short, memorable sections to make revising easier.

Revised

4–6 weeks to go

- Read through the relevant sections of this book and refer to the examiners' tips, examiners' summaries, typical mistakes and key terms. Tick off the topics as you feel confident about them. Highlight those topics you find difficult and look at them again in detail.
- Test your understanding of each topic by working through the 'Now test yourself' questions in the book. Look up the answers at the back of the book.
- Make a note of any problem areas as you revise, and ask your teacher to go over these in class.
- Look at past papers. They are one of the best ways to revise and practise your exam skills. Write or prepare planned answers to the exam practice questions provided in this book. Check your answers online and try out the extra quick quizzes at **www.therevisionbutton.co.uk/myrevisionnotes**
- Use the revision activities to try out different revision methods. For example, you can make notes using mind maps, spider diagrams or flash cards.
- Track your progress using the revision planner and give yourself a reward when you have achieved your target.

Revised

One week to go

- Try to fit in at least one more timed practice of an entire past paper and seek feedback from your teacher, comparing your work closely with the mark scheme.
- Check the revision planner to make sure you haven't missed out any topics. Brush up on any areas of difficulty by talking them over with a friend or getting help from your teacher.
- Attend any revision classes put on by your teacher. Remember, he or she is an expert at preparing people for examinations.

Revised

The day before the examination

- Flick through these Revision Notes for useful reminders, for example the Examiners' tips, examiners' summaries, typical mistakes and key terms.
- Check the time and place of your examination.
- Make sure you have everything you need — extra pens and pencils, tissues, a watch, bottled water, sweets.
- Allow some time to relax and have an early night to ensure you are fresh and alert for the examination.

Revised

My exams

A2 Chemistry Unit 4

Date: ....

Time: ....

Location:..

A2 Chemistry Unit 5

Date: ....

Time: ....

Location:..

1 Kinetics

Rate equations

What is meant by 'rate of a chemical reaction'?

Revised

For the reaction:

$A + B \rightarrow X + Y$

the **rate of reaction** is equal to the rate at which the concentration of reactant A or B is decreasing, or the rate at which the concentration of product X or Y is increasing.

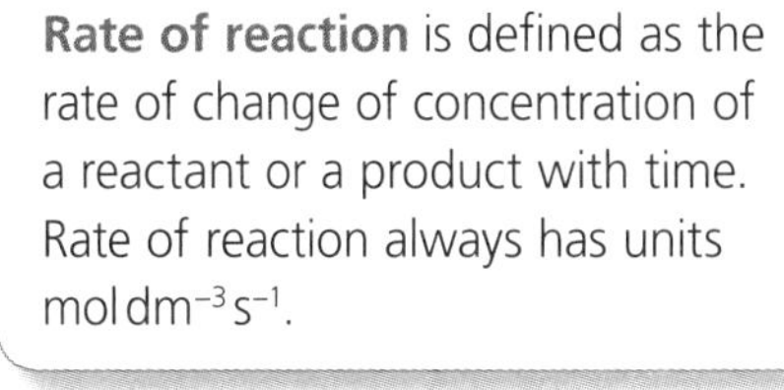

Rate of reaction is defined as the rate of change of concentration of a reactant or a product with time. Rate of reaction always has units $mol\,dm^{-3}\,s^{-1}$.

The rate of a reaction can be measured at any time during the reaction, but it is most conveniently done at the very start of the reaction, at time $t = 0$. This rate is called the **initial rate**. The initial rate of reaction is calculated by drawing the **tangent** to the concentration–time graph at $t = 0$.

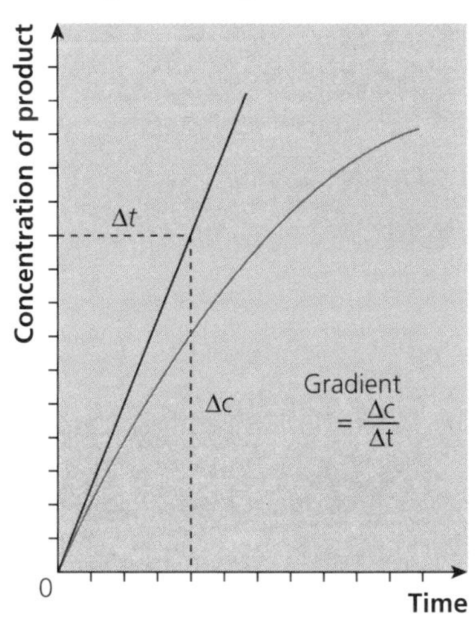

Figure 1.1 Calculating the initial rate of reaction

It is possible to measure the rate at any time during the reaction by measuring the gradient of the tangent at that particular time.

Concentration and rate

Revised

For the general reaction between reactants A and B in solution:

$A(aq) + B(aq) \rightarrow X + Y$

$\text{rate} = k[A]^a[B]^b$

where square brackets, [ ], represent the molar concentrations.

The variables a and b are called the **orders** of reaction, with respect to each reactant, and k is called the **rate constant**.

The rate constant is constant with time and varying concentration, but it does increase with increasing temperature.

In the equation above, the order with respect to A is *a*, and the order with respect to B is *b*. Orders will only have the (integer) values 0, 1 and 2 in examples at this level.

Example

For the reaction in aqueous solution between propanone, $CH_3COCH_3(aq)$, and iodine, $I_2(aq)$, in the presence of an acid catalyst:

$$CH_3COCH_3 + I_2 \rightarrow CH_3COCH_2I + HI$$

the rate equation is found by experiment to be:

$$\text{rate} = k[CH_3COCH_3]^1[I_2]^0[H^+]^1$$

So the rate of this reaction is **first order** with respect to propanone, **first order** with respect to hydrogen ions, and **zero order** with respect to iodine.

Typical mistake

Many candidates think that a rate equation can be deduced from the balanced symbol equation. This is not true. It has to be determined experimentally.

The **overall order** of a reaction is simply the sum of all of the individual orders. If the rate equation is:

$$\text{rate} = k[CH_3COCH_3]^1[I_2]^0[H^+]^1$$

the overall order is 1 + 0 + 1 = 2, or **second order** overall.

The **units of the rate constant** can be found by rearranging the rate equation. For the propanone/iodine reaction above:

$$\text{rate} = k[CH_3COCH_3]^1[I_2]^0[H^+]^1$$

Making *k* the subject of this equation gives:

$$k = \frac{\text{rate}}{[CH_3COCH_3(aq)]^1[I_2(aq)]^0[H^+(aq)]^1}$$

So the units of *k* are $\frac{\text{mol dm}^{-3}\,\text{s}^{-1}}{(\text{mol dm}^{-3})^2}$

$= \text{mol}^{-1}\,\text{dm}^3\,\text{s}^{-1}$

Examiners' tip

Remember the power rules from your maths lessons:

$$(y^a)\times(y^b) = y^{(a+b)} \text{ and } \frac{y^a}{y^b} = y^{(a-b)} \text{ and } (y^a)^b = y^{ab}$$

Now test yourself

1 The rate equation for the reaction:

$$(CH_3)_3CBr + OH^- \rightarrow (CH_3)_3COH + Br^-$$

is found to be rate = $k[(CH_3)_3CBr]$. The rate is measured in $\text{mol dm}^{-3}\,\text{s}^{-1}$.

(a) What are the orders with respect to $(CH_3)_3CBr$ and OH^-?

(b) What is the overall order of the reaction?

(c) Give the units of the rate constant.

Answers on p. 108

Tested

Order of reaction and rate

Revised

When the concentration of a reactant in a chemical reaction is monitored with time, it may vary in one of three main ways, as indicated by the graphs shown in Figure 1.2.

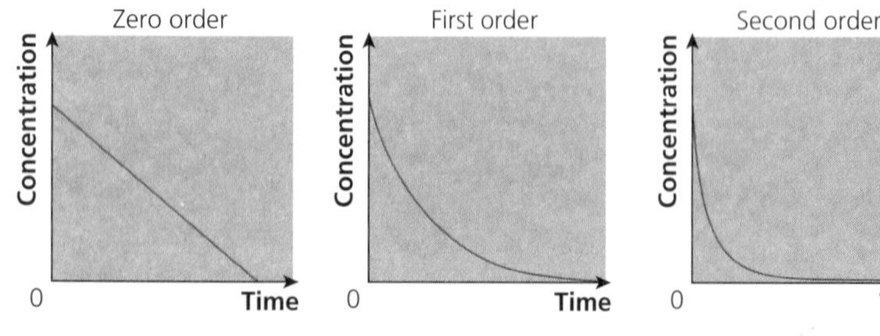

Figure 1.2 Concentration–time graphs

When the rates of reaction are monitored against concentration, the graphs shown in Figure 1.3 are obtained.

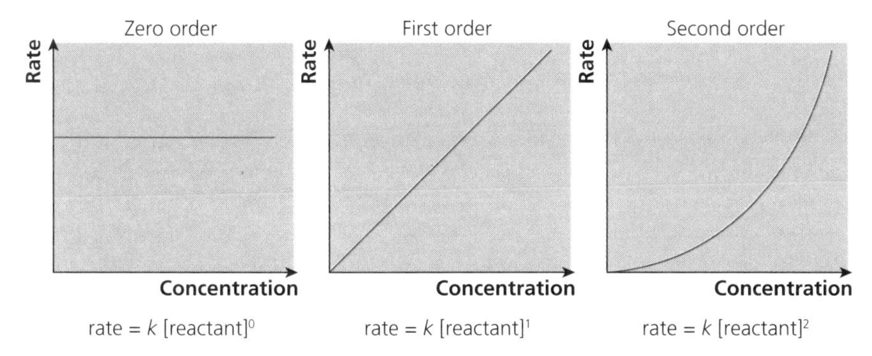

Figure 1.3 Rate–concentration graphs

- Zero order — when the concentration is doubled, the rate is unaffected.
- First order — when the concentration is doubled, the rate increases by a factor of 2.
- Second order — when the concentration is doubled, the rate increases by a factor of 4.

Examiners' tip

Remember — 'first order' means that when the concentration of the reactant is multiplied by a factor *x*, the rate is also multiplied by the factor *x*.

Determining rate equations

In a reaction between reactants A, B and C, the following initial rates were obtained using the initial concentrations shown in four different experiments.

Experiment	Initial concentration/mol dm^{-3}			Initial rate/mol dm^{-3} s^{-1}
	[A]	[B]	[C]	
1	0.05	0.03	0.12	1.20×10^{-4}
2	0.20	0.06	0.24	3.84×10^{-3}
3	0.05	0.03	0.24	2.40×10^{-4}
4	0.20	0.03	0.12	1.92×10^{-3}

Comparing experiments 1 and 3:

[A] and [B] remain the same but [C] doubles. This doubles the rate so the order with respect to C must be 1.

Comparing experiments 1 and 4:

[A] is multiplied by 4 and [B] and [C] stay constant. This multiplies the rate by a factor of 16. The order with respect to A must therefore be 2, because $4^2 = 16$.

Comparing experiments 3 and 2:

[A] is multiplied by 4, [B] is multiplied by 2 and [C] stays constant. This multiplies the rate by a factor of 16. We know that the order with respect to A is 2, so multiplying [A] by 4 should multiply the rate by 16 — and it does. Therefore the change in [B] has no effect on the rate. So the order with respect to B must be 0.

The overall rate equation is therefore:

$$\text{rate} = k[A]^2[B]^0[C]^1$$

and the reaction is third order overall.

Examiners' tip

Questions asking you to deduce a rate equation using concentration rate data are very common in the exam.

The value for the rate constant can be calculated using any of experiments 1, 2, 3 or 4. Using the data from experiment 1 and that rate = $k[A]^2[C]$ gives:

$$1.2 \times 10^{-4} = k \times 0.05^2 \times 0.12^1$$

This gives k = 0.4, with units $\frac{\text{mol dm}^{-3}\,\text{s}^{-1}}{(\text{mol dm}^{-3})^3} = \text{mol}^{-2}\,\text{dm}^6\,\text{s}^{-1}$

Now test yourself

Tested

2 The following data were measured for the reaction between nitrogen(II) oxide and hydrogen:

$$2NO(g) + 2H_2(g) \rightarrow N_2(g) + 2H_2O(g)$$

Experiment number	Initial concentrations/mol dm^{-3}		Initial rate/ mol dm^{-3}s^{-1}
	[NO(g)]	[H_2(g)]	
1	0.100	0.100	1.11×10^{-3}
2	0.100	0.200	2.24×10^{-3}
3	0.200	0.100	4.45×10^{-3}

(a) Determine the order with respect to NO and the order with respect to H_2.
(b) What is the overall order of reaction?
(c) Write the rate equation for the reaction.
(d) Determine a value for the rate constant and deduce its units.

Answers on p. 108

The effect of temperature on rate

In all chemical reactions, as the temperature is increased the rate will increase.

A typical Maxwell–Boltzmann distribution is shown in Figure 1.4.

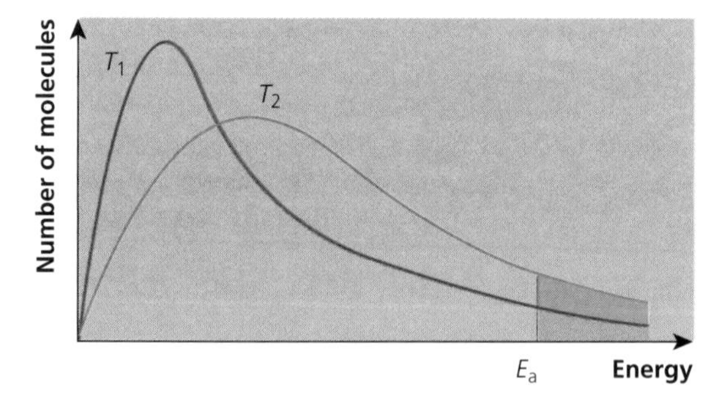

Figure 1.4 Graph of a Maxwell–Boltzmann distribution at two different temperatures, where $T_2 > T_1$

The rate increases with increasing temperature because the proportion of particles having energy greater than the activation energy increases. This increases the number of successful collisions taking place per unit time.

Given a typical rate equation:

rate = $k[A]^2[C]$

as the temperature increases, the value of k will also increase.

- The rate constant is related to the shaded area of the Maxwell–Boltzmann distribution in Figure 1.4 to the right of the activation energy value. The larger this area, the more particles have energy $E > E_a$ so both the rate and the rate constant will be larger.
- The lower the temperature, the smaller will be the proportion of molecules having an energy greater than the activation energy. Therefore, the rate constant is smaller.

Rate equations and reaction mechanism

A rate equation tells us how the concentrations of the reactants affect the rate of the reaction. However, it also indicates something about the **mechanism** of the reaction taking place.

For example, nitrogen(IV) oxide reacts with fluorine according to the equation:

$$2NO_2(g) + F_2(g) \rightarrow 2NO_2F(g)$$

The rate equation, found by experiment, is first order with respect to both reactants:

$$\text{rate} = k[NO_2(g)][F_2(g)]$$

We can deduce the following about the mechanism of the reaction:

- Since both NO_2 and F_2 are in the rate equation as non-zero orders, this means that both their concentrations affect the rate.
- Any species that is zero order in the mechanism must be involved in a subsequent fast step.

Therefore, it can be deduced that:

- NO_2 and F_2 are involved in the **rate-determining step** of the mechanism — so the first step could be:

  $NO_2(g) + F_2(g) \rightarrow NO_2F(g) + F(g)$ slow
- The next step — a fast one — may involve F(g) from the slow step reacting with another NO_2 molecule:

  $NO_2(g) + F(g) \rightarrow NO_2F(g)$ fast

The species appearing in the rate equation also occur in the **rate-determining step** of the mechanism.

Notice how the two individual steps in this mechanism add up to give the overall chemical equation:

Step 1: $NO_2(g) + F_2(g) \rightarrow NO_2F(g) + F(g)$ slow

Step 2: $NO_2(g) + F(g) \rightarrow NO_2F(g)$ fast

Overall: $2NO_2(g) + F_2(g) \rightarrow 2NO_2F(g)$

Exam practice

1 In a reaction taking place between substances A and B, the following results were obtained in three different experiments.

Assume that the concentrations given in the table are all initial concentrations.

	[A]/mol dm^{-3}	[B]/mol dm^{-3}	Initial rate/mol dm^{-3} s^{-1}
Experiment 1	1.0×10^{-2}	4.0×10^{-3}	3.20×10^{-3}
Experiment 2	1.0×10^{-2}	8.0×10^{-3}	1.28×10^{-2}
Experiment 3	2.0×10^{-2}	8.0×10^{-3}	2.56×10^{-2}

(a) Deduce the order of the reaction with respect to A, and the order of the reaction with respect to B. [2]

(b) Hence, write the rate equation for the reaction and calculate the rate constant stating its units. [3]

(c) Calculate the rate of reaction when the initial concentrations of A and B are both 6.0×10^{-3} mol dm^{-3}. [1]

(d) In another experiment in which the initial concentrations of A and B are both x mol dm^{-3}, the initial rate of reaction is found to be 9.22×10^{-2} mol dm^{-3} s^{-1}. Calculate the value of x. [2]

2 Propanone, in acidic solution, reacts with iodine according to:

$CH_3COCH_3 + I_2 \rightarrow CH_3COCH_2I + HI$

In an experiment, the time taken for the iodine to reach a certain concentration was measured. The concentration of hydrogen ions was kept constant in all four experiments described in the table below. The order with respect to the acid is known to be 1.

[CH_3COCH_3]/mol dm^{-3}	[I_2]/mol dm^{-3}	Time/s
0.25	0.05	68
0.50	0.05	34
1.00	0.05	17
0.50	0.10	34

(a) Deduce the order of reaction with respect to propanone. [1]

(b) Deduce the order with respect to iodine. [1]

(c) Hence, write the rate equation for the reaction. [1]

(d) Comment, with a reason, whether or not the following mechanism is consistent with the rate equation you have suggested in part (c). [2]

Step 1: $CH_3COCH_3 + H^+ \rightarrow [CH_3C(OH)CH_3]^+$ slow

Step 2: $[CH_3C(OH)CH_3]^+ \rightarrow CH_3C(OH)CH_2 + H^+$ fast

Step 3: $CH_3C(OH)CH_2 + I_2 \rightarrow CH_3COCH_2I + HI$ fast

Answers and quick quizzes online

Online

Examiners' summary

You should now have an understanding of:

- ✔ what is meant by rate of reaction and initial rate
- ✔ the effect of changes in concentration on rate
- ✔ the rate constant and how to determine its units
- ✔ rate equations and how to determine them given initial rate information
- ✔ concentration–time and rate–concentration graphs
- ✔ the effect of temperature on reaction rate
- ✔ the link between a rate equation and a reaction mechanism

2 Equilibria

The equilibrium constant, K_c

For all reactions that form equilibrium mixtures, there is a relationship between the concentration of the reactants and the concentration of the products measured at equilibrium. This is stated in the **equilibrium law**.

In the general reaction:

$$aA + bB \rightleftharpoons cC + dD$$

the **equilibrium constant** is given by:

$$K_c = \frac{[C]^c \times [D]^d}{[A]^a \times [B]^b}$$

- K_c is a constant and is not affected by changes in concentration, pressure, surface area or the presence of a catalyst. Only changes in temperature affect the value of K_c.
- The larger the value of K_c the greater is the proportion of the products compared to the reactants at equilibrium. If K_c is small the reactants predominate, and if K_c is large then the products predominate in the equilibrium mixture.

Units of an equilibrium constant

Revised

The units of K_c will depend on the expression used. For example, for the reaction:

$$N_2O_4(g) \rightleftharpoons 2NO_2(g)$$

$$K_c = \frac{[NO_2(g)]^2}{[N_2O_4(g)]}$$

So, the units of K_c will be $\frac{(\text{mol dm}^{-3})^2}{\text{mol dm}^{-3}} = \text{mol dm}^{-3}$

The units for K_c for the reaction:

$$2SO_2(g) + O_2(g) \rightleftharpoons 2SO_3(g)$$

are given by:

$$K_c = \frac{[SO_3(g)]^2}{[SO_2(g)]^2[O_2(g)]}$$

$$= \frac{(\text{mol dm}^{-3})^2}{(\text{mol dm}^{-3})^2 \times \text{mol dm}^{-3}} = \text{mol}^{-1}\,\text{dm}^3$$

Now test yourself

1 Write equilibrium expressions for these two reactions:

(a) $H_2(g) + I_2(g) \rightleftharpoons 2HI(g)$

(b) $PCl_5(g) \rightleftharpoons PCl_3(g) + Cl_2(g)$

(c) What are the units of the equilibrium constants in parts (a) and (b)?

Answers on p. 108

Tested

Calculations involving equilibrium constants

Revised

Example 1

Calculate the value of K_c for the following reaction in the gas phase:

$$N_2(g) + 3H_2(g) \rightleftharpoons 2NH_3(g)$$

given that the concentrations (in $mol\,dm^{-3}$) of nitrogen, hydrogen and ammonia measured at equilibrium are 0.0220, 0.0460 and 0.0560 respectively.

Answer

$$K_c = \frac{[NH_3]^2}{[N_2][H_2]^3}$$

$$= \frac{(0.0560)^2}{0.0220 \times (0.0460)^3}$$

$$= 1.46 \times 10^3\,mol^{-2}\,dm^6$$

Examiners' tip

Calculations involving K_c can look tricky — make sure you practise a range of them because there are few different types of questions that will be asked.

Example 2

3.00 moles of ethanol and 3.00 moles of ethanoic acid were mixed and the mixture was allowed to reach equilibrium. It was found that there were 1.40 moles of ethyl ethanoate, $CH_3COOC_2H_5$, in the equilibrium mixture. Calculate K_c for the reaction.

$$CH_3COOH(l) + CH_3CH_2OH(l) \rightleftharpoons CH_3COOC_2H_5(l) + H_2O(l)$$

Answer

$$K_c = \frac{[CH_3COOC_2H_5(l)][H_2O(l)]}{[CH_3COOH(l)][CH_3CH_2OH(l)]}$$

Writing the number of moles of each substance underneath each substance in the reaction:

	$CH_3COOH(l)$ +	$CH_3CH_2OH(l)$ ⇌	$CH_3COOC_2H_5(l)$ +	$H_2O(l)$
At start/mol:	3.00	3.00	0	0

Let x be the number of moles of ethanoic acid that reacted:

At equilibrium/mol:	$3.00 - x$	$3.00 - x$	x	x

But the amount of ethyl ethanoate at equilibrium is 1.40 mol, so $x = 1.40$.

We can now deduce the amount of each of the substances present in the equilibrium mixture:

- $CH_3COOH(l)$: $3.00 - x = 3.00 - 1.40 = 1.60$ mol
- $CH_3CH_2OH(l)$: $3.00 - x = 3.00 - 1.40 = 1.60$ mol
- $CH_3COOC_2H_5(l)$: $x = 1.40$ mol
- $H_2O(l)$: $x = 1.40$ mol

Substituting into the expression for K_c for this reaction and letting V_{tot} be the total volume, gives:

$$K_c = \frac{(1.40/V_{tot}) \times (1.40/V_{tot})}{(1.60/V_{tot}) \times (1.60/V_{tot})}$$

The volume terms cancel to give:

$$K_c = \frac{1.40^2}{1.60^2}$$

= 0.766 (no units)

Example 3

N_2O_4 dissociates on heating to form an equilibrium mixture according to the equation:

$$N_2O_4(g) \rightleftharpoons 2NO_2(g)$$

1.00 mol of N_2O_4 is allowed to reach equilibrium and it is found that 0.24 mol of N_2O_4 remains in the mixture. The total volume of the container is 2.00 dm³. Calculate K_c for the reaction.

Answer

$$K_c = \frac{[NO_2(g)]^2}{[N_2O_4(g)]}$$

Let x be the number of moles of N_2O_4 that react.

$$N_2O_4(g) \rightleftharpoons 2NO_2(g)$$

At start/mol: 1.00 0

The 1:2 mole ratio in the equation shows that when x mol of N_2O_4 reacts, then $2x$ mol of NO_2 is formed.

At equilibrium/mol: $1.00 - x$ $2x$

But $1.00 - x = 0.24$, so $x = 0.76$ mol. Substituting gives:

At equilibrium/mol 0.24 2×0.76

The concentrations, in mol dm⁻³, are given by:

$\frac{0.24}{2.00}$ $\frac{1.52}{2.00}$

$0.12\ \text{mol dm}^{-3}$ $0.76\ \text{mol dm}^{-3}$

$$K_c = \frac{(0.76)^2}{0.12}$$

$= 4.81\ \text{mol dm}^{-3}$

Typical mistake

The major error in doing these calculations is using the number of moles rather than the concentration when substituting into the K_c expression. Unless the volume terms cancel you will have to divide the molar amounts by the mixture volume.

Now test yourself

2 In the equilibrium:

$$PCl_5(g) \rightleftharpoons PCl_3(g) + Cl_2(g)$$

0.550 mol of PCl_5 is allowed to dissociate at 500 K in a total volume of 2.00 dm³. It was found that 0.240 mol of PCl_5 remained at equilibrium. Calculate K_c.

Answers on p. 108

Tested

The effect of temperature, concentration and catalysts on K_c

Temperature

Revised

Table 2.1 shows how the value of the equilibrium constant changes with **temperature** for the reaction of hydrogen with iodine:

$$H_2(g) + I_2(g) \rightleftharpoons 2HI(g)$$

As the temperature increases:

- the value of K_c decreases
- the equilibrium position must be shifting towards the left-hand side
- more of the reactants and less of the products are forming.

Temperature has an effect on the value of an equilibrium constant. Whether the value of K_c increases or decreases with temperature depends on the sign of the enthalpy change for the reaction.

Table 2.1 Changes in K_c values with temperature

Temperature/K	K_c
300	800
500	150
700	55
770	25

The forward reaction must be **exothermic** — according to Le Chatelier's principle — an increase in temperature favours formation of the *reactants* in an exothermic process. This is because the reverse endothermic reaction is favoured to oppose the increase in temperature.

Examiners' tip

Temperature is the only factor that affects K_c. All other variables — concentration, pressure, surface area, addition of a catalyst — have no effect on K_c.

Concentration

Revised

If a reaction is at equilibrium and changes are made by increasing the amount of one of the substances present, the equilibrium position shifts so as to reduce the effect of this change. When the new **equilibrium concentrations** are used to calculate K_c the value is found to be the same as it was before (assuming the temperature remains unchanged).

Concentration changes do not affect the value of an equilibrium constant.

Catalysts

Revised

A **catalyst** works by providing an alternative route for a reaction with a lower activation energy — it will make the reaction go faster, or the rate at which equilibrium is attained will be higher. Both the forward and the reverse reactions are faster. Therefore equilibrium is reached sooner, but the position of equilibrium is unchanged. A catalyst has no effect on the composition of the equilibrium mixture.

Catalysts have no effect on the value of K_c — they only affect the rate of the reaction.

Exam practice

1 A gaseous equilibrium is established by adding 2.00 mol of sulfur(IV) oxide to 2.00 mol of oxygen and waiting for equilibrium to be attained:

$$2SO_2(g) + O_2(g) \rightleftharpoons 2SO_3(g)$$

It is found that 0.550 mol of sulfur(VI) oxide exists in the equilibrium mixture. The total volume of the reaction mixture is 10.0 dm^3.

(a) Calculate the amounts, in moles, of SO_2 and O_2 at equilibrium. [2]

(b) Calculate the concentrations of all species taking part in the reaction in mol dm^{-3}. [3]

(c) Write an expression for K_c for the reaction. [1]

(d) Hence determine the value of K_c for the equilibrium stating the units. [3]

Sulfur dioxide, SO_2, is added to the equilibrium, and the pressure is increased at constant volume (maintaining the same temperature).

(e) State and explain the effect on:

(i) the position of equilibrium [2]

(ii) the value of K_c. [2]

Answers and quick quizzes online

Online

Now test yourself

3 Given the general equilibrium:

$$A(aq) \rightleftharpoons B(aq) + C(aq)$$

it is found that the values for K_c vary with increasing temperature as follows:

Temperature/K	K_c/mol dm^{-3}
200	0.00140
250	0.120
300	1.56
350	5.76

(a) Write an expression for K_c for the reaction, stating the units.

(b) Explain whether the reaction is exothermic or endothermic.

(c) A catalyst is added to the reaction mixture prior to it reaching equilibrium. Explain what effect this will have on the value of K_c.

Answers on p. 108

Tested

Examiners' summary

You should now have an understanding of:

- ✔ an equilibrium constant, K_c
- ✔ the qualitative effects of changes of temperature and concentration in terms of the position of equilibrium and on K_c

3 Acid–base equilibria

Brønsted–Lowry acids and bases

Reactions between acids and bases in solution involve the **transfer of a proton** from one species to another — for example, in the reaction between hydroiodic acid, HI, and ethanoic acid, CH_3COOH:

$$HI(aq) + CH_3COOH(aq) \rightleftharpoons I^-(aq) + CH_3COOH_2^+(aq)$$

it can be seen that:

- HI has donated a proton to CH_3COOH, so HI is a **Brønsted–Lowry acid**
- CH_3COOH has accepted a proton from HI, so CH_3COOH is a **Brønsted–Lowry base** in this reaction.

It is possible to pair up each species with the one it forms to make a conjugate acid–base pair. For example, HI is a Brønsted–Lowry acid in the forward direction, but I^- is a base in the reverse direction:

$$HI(aq) + CH_3COOH(aq) \rightleftharpoons I^-(aq) + CH_3COOH_2^+(aq)$$

acid 1 base 2 base 1 acid 2

A **Brønsted–Lowry acid** is a proton donor; a **Brønsted–Lowry base** is a proton acceptor.

Examiners' tip

The difference between an acid and its conjugate base (or vice versa) is simply one proton — for example HI (acid) and I^- (conjugate base) or CH_3COOH (base) and $CH_3COOH_2^+$ (conjugate acid).

Strong and weak acids and bases

Hydrochloric acid, HCl(aq), undergoes complete dissociation in aqueous solution — it is a **strong acid**. In other words, virtually all of the **molecular** hydrogen chloride, HCl, once dissolved in water, dissociates completely to form **ions**:

$$HCl(aq) + H_2O(l) \rightarrow H_3O^+(aq) + Cl^-(aq)$$

Strong bases like sodium hydroxide dissolve readily in water and are fully ionised forming hydroxide ions, $OH^-(aq)$. A soluble base is called an **alkali**.

$$NaOH(s) + aq \rightarrow Na^+(aq) + OH^-(aq)$$

Weak acids and **weak bases** interact with water molecules to form equilibrium mixtures.

The weak acid ethanoic acid, CH_3COOH, dissociates in water as follows:

$$CH_3COOH(aq) + H_2O(l) \rightleftharpoons CH_3COO^-(aq) + H_3O^+(aq)$$

The weak base ammonia, NH_3, dissociates in water as follows:

$$NH_3(aq) + H_2O(l) \rightleftharpoons NH_4^+(aq) + OH^-(aq)$$

Strong acids and **strong bases** dissociate completely in solution.

Weak acids and **weak bases** dissociate partially in solution.

Now test yourself

1 Label the conjugate acid–base pairs in the reaction between sulfuric(VI) acid and nitric(V) acid:

$$HNO_3(aq) + H_2SO_4(aq) \rightleftharpoons HSO_4^-(aq) + H_2NO_3^+(aq)$$

2 Write equations to show the dissociation of the following in aqueous solution:

(a) benzoic acid, C_6H_5COOH — a weak acid

(b) methylamine, CH_3NH_2 — a weak base

Answers on p. 108

Tested

pH

This is a measure of the 'acidity' (or 'alkalinity') of an aqueous solution. pH is defined mathematically:

$$pH = -\log_{10}[H^+(aq)]$$

Example 1

Calculate the pH of 0.00560 $mol\,dm^{-3}$ HCl(aq).

Answer

Hydrochloric acid is a strong acid and so we can assume complete dissociation.

$HCl(aq) + H_2O(l) \rightarrow H_3O^+(aq) + Cl^-(aq)$ or
$HCl(aq) \rightarrow H^+(aq) + Cl^-(aq)$

So the concentration of hydrogen ions, $H^+(aq)$, will be 0.00560 $mol\,dm^{-3}$.

$pH = -\log_{10} [H^+(aq)]$
$= -\log_{10}(0.00560)$
$= -(-2.25) = 2.25$

Examiners' tip

Make sure that you quote pH to 2 decimal places in your answers.

Example 2

Calculate the pH of 0.100 $mol\,dm^{-3}$ dilute sulfuric(VI) acid. Assume that the acid is completely dissociated.

Answer

$H_2SO_4(aq) + 2H_2O(l) \rightarrow 2H_3O^+(aq) + SO_4^{2-}(aq)$ or
$H_2SO_4(aq) \rightarrow 2H^+(aq) + SO_4^{2-}(aq)$

So the concentration of hydrogen ions, $H^+(aq)$, will be (0.100 × 2) $mol\,dm^{-3}$.

$pH = -\log_{10} [H^+(aq)]$
$= -\log_{10}(0.200)$
$= 0.70$

Calculating the concentration of hydrogen ions

Revised

Because $pH = -\log_{10} [H^+(aq)]$, rearranging this to make $[H^+(aq)]$ the subject gives:

$[H^+(aq)] = 10^{-pH}$

Example 1

Calculate the concentration of hydrogen ions in a solution that has a pH of 13.20.

Answer

$[H^+(aq)] = 10^{-13.20}$
$= 6.31 \times 10^{-14}\ mol\,dm^{-3}$

The ionic product of water, K_w

Water dissociates very slightly according to the equilibrium:

$H_2O(l) \rightleftharpoons H^+(aq) + OH^-(aq)$

An equilibrium constant (similar to K_c) can be written for this dissociation. It has the symbol K_w and is called the **ionic product of water**:

$K_w = [H^+(aq)][OH^-(aq)]$

The value of K_w is $1.00 \times 10^{-14}\ mol^2\,dm^{-6}$ at 298 K.

The value of K_w depends, like all equilibrium constants, on temperature. As the temperature increases, so does K_w. This indicates that the dissociation of water is an endothermic process. As the temperature increases, the concentration of $H_3O^+(aq)$ and $OH^-(aq)$ also increase.

Calculating the pH of solutions

Revised

The pH of a solution of a strong base can be calculated using the ionic product of water, K_w

Example 1

Calculate the pH of 9.40×10^{-3} mol dm^{-3} sodium hydroxide solution, NaOH(aq), at 298 K. The value of K_w is 1.00×10^{-14} mol^2 dm^{-6} at 298 K.

Answer

$[OH^-(aq)] = 9.40 \times 10^{-3}$ mol dm^{-3}

$K_w = [H^+(aq)][OH^-(aq)]$

$1.00 \times 10^{-14} = [H^+(aq)] \times (9.40 \times 10^{-3})$

$[H^+(aq)] = 1.064 \times 10^{-12}$ mol dm^{-3}

$pH = -\log_{10}[H^+(aq)]$

$= -\log_{10}(1.064 \times 10^{-12})$

$= 11.97$

Example 2

Calculate the pH of the mixture formed when 20.0 cm^3 of 0.200 mol dm^{-3} H_2SO_4 is added to 40.0 cm^3 of 0.250 mol dm^{-3} NaOH.

Answer

Moles of $H_2SO_4 = \frac{20.0}{1000} \times 0.200$

$= 4.00 \times 10^{-3}$ mol

Moles of NaOH $= \frac{40.0}{1000} \times 0.250$

$= 1.00 \times 10^{-2}$ mol

The equation for the reaction is:

$$2NaOH + H_2SO_4 \rightarrow Na_2SO_4 + 2H_2O$$

4.00×10^{-3} moles of H_2SO_4 will react exactly with 8.00×10^{-3} moles of NaOH (using the 1:2 ratio from the equation).

However, 1.00×10^{-2} moles of NaOH were originally present, so $(1.00 \times 10^{-2} - 8.00 \times 10^{-3})$ mol of NaOH will remain — that is 2.00×10^{-3} mol. This is present in a total volume of 60.0 cm^3.

$[OH^-] = \frac{1000}{60.0} \times 2.00 \times 10^{-3}$

$= 0.0333$ mol dm^{-3}

But $K_w = [H^+][OH^-] = 1.00 \times 10^{-14}$

So $[H^+] = \frac{K_w}{[OH^-]} = \frac{1.00 \times 10^{-14}}{0.0333}$

$= 3.00 \times 10^{-13}$ mol dm^{-3}

$pH = -\log_{10}[H^+(aq)]$

$= -\log_{10}(3.00 \times 10^{-13})$

$= 12.52$

Now test yourself

3 Calculate the pH of the following solutions of strong acids and bases at 298 K.

Assume the value of K_w is 1.00×10^{-14} mol^2 dm^{-6}

(a) 6.60×10^{-2} mol dm^{-3} HNO_3(aq)

(b) 5.67×10^{-4} mol dm^{-3} NaOH(aq)

(c) 0.0500 mol dm^{-3} KOH(aq)

(d) 0.0950 mol dm^{-3} H_2SO_4(aq).

4 A sample of rain water is found to have a pH of 6.45. What is the concentration of hydrogen ions and the concentration of hydroxide ions in the rain water at 298 K?

Answers on p. 108

Tested

The acid dissociation constant, K_a

A weak acid, HA, dissociates in water according to the equilibrium:

$HA(aq) + H_2O(l) \rightleftharpoons H_3O^+(aq) + A^-(aq)$ or

$HA(aq) \rightleftharpoons H^+(aq) + A^-(aq)$

and an equilibrium constant, K_a, can be written:

$$K_a = \frac{[H_3O^+(aq)][A^-(aq)]}{[HA(aq)]} \text{ or } \frac{[H^+][A^-]}{[HA]}$$

K_a has units of $mol\,dm^{-3}$.

It is also possible to convert K_a values into pK_a values:

$pK_a = -\log_{10}K_a$

$K_a = 10^{-pK_a}$

As acids increase in strength, the values of K_a increase and the values of pK_a decrease (or stronger acids have larger K_a values and smaller pK_a values).

The pK_a values in Table 3.1 show acids that decrease in strength from top to bottom.

Table 3.1 Typical pK_a values

Hydrofluoric acid, HF	3.20
Benzoic acid, C_6H_5COOH	4.20
Ethanoic acid, CH_3COOH	4.76
Hydrogen cyanide, HCN	9.40

Typical mistake

Many candidates include water in an expression for K_a. Water is never put into a K_a expression because its concentration is taken as constant.

Typical mistake

Many candidates state wrongly that acidic strength increases as pK_a increases. Remember that acidic strength increases as K_a increases, but as pK_a decreases.

Calculating the pH of a solution of a weak acid

Revised

The pH of a solution of a weak acid is calculated in a similar way to that for a solution of a strong acid — with an adjustment for the fact that incomplete dissociation occurs.

Example 1

Calculate the pH of a solution of butanoic acid, $CH_3(CH_2)_2COOH$, of concentration $0.150\,mol\,dm^{-3}$. K_a for butanoic acid is $1.514 \times 10^{-5}\,mol\,dm^{-3}$ at 298 K.

Answer

	$CH_3(CH_2)_2COOH(aq)$	$\rightleftharpoons H_3O^+(aq)$	$+ CH_3(CH_2)_2COO^-(aq)$
At start/mol:	0.150	0	0

Let x be the number of moles of acid that dissociate in $1.0\,dm^3$ of solution:

At equilibrium/mol:	$0.150 - x$	x	x

Because the acid is weak, we can make the approximation that 0.150 is much larger than x, and can ignore x in the left-hand term.

Concentration at equilibrium/$mol\,dm^{-3}$:

$\frac{0.150}{1}$	$\frac{x}{1}$	$\frac{x}{1}$

$$K_a = \frac{[CH_3(CH_2)_2COO^-(aq)][H^+(aq)]}{[CH_3(CH_2)_2COOH(aq)]}$$

Substituting the values we know:

$$1.514 \times 10^{-5} = \frac{x^2}{0.150}$$

$$x^2 = 1.514 \times 10^{-5} \times 0.150$$

$$x = \sqrt{1.514 \times 10^{-5} \times 0.150}$$

$= 1.507 \times 10^{-3}\,\text{mol dm}^{-3}$

This is equal to the hydrogen ion concentration, $[H_3O^+(aq)]$, so

$pH = -\log_{10}[H^+(aq)]$

$= -\log_{10}(1.507 \times 10^{-3})$

$= 2.82$

Examiners' tip

It looks a long calculation, but with practice you will realise that these are not that difficult.

It is acceptable to avoid some of the algebra in the method above and use:

$$K_a = \frac{[H^+(aq)][A^-(aq)]}{[HA(aq)]}$$

For a solution of weak acid $[H^+] = [A^-]$ so

$$K_a = \frac{[H^+]^2}{[HA]}$$

Therefore, $[H^+]^2 = K_a \times [HA]$

$$[H^+(aq)] = \sqrt{K_a \times [HA(aq)]}$$

Now test yourself Tested

5 **(a)** Write an equation to show the partial dissociation of ethanoic acid, CH_3COOH, in water.

(b) Write an expression for the acid dissociation constant for ethanoic acid.

(c) Calculate the pH of a $0.100\,\text{mol dm}^{-3}$ aqueous solution of ethanoic acid.

$[K_a$ for ethanoic acid is $1.74 \times 10^{-5}\,\text{mol dm}^{-3}]$

Answers on p. 108

Titrations

A titration curve is produced when a base is titrated against an acid and the pH is monitored as the volume of added base increases.

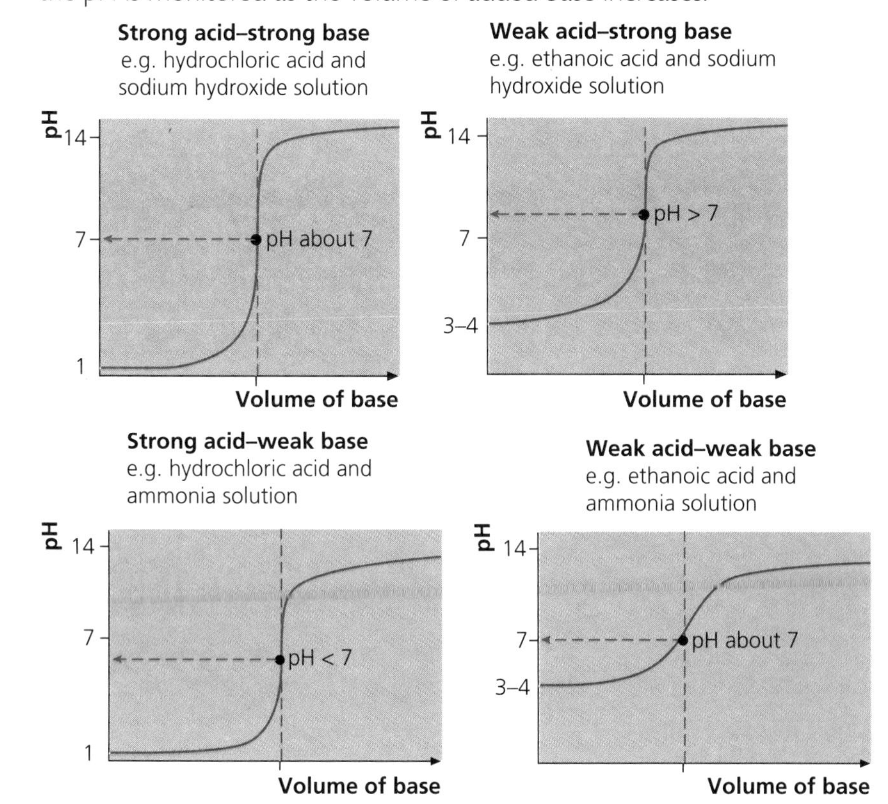

Figure 3.1 Titration curves

The shapes of these curves depend on many factors.

- The starting and final pH values depend on the strength of the acid and/or base used. For example, if hydrochloric acid (a strong acid) is used then the starting pH will be approximately equal to 1; if ammonia solution (a weak base) is used then a final pH of 9–10 will result.
- The steep part of the curve, where there is a rapid change of pH, shows the volume of base needed for **equivalence** with the amount of acid used.
- The equivalence pH depends on the strengths of the acid and base used and the nature of the salt formed at the end-point — Table 3.2 shows some typical cases.

Table 3.2 Typical titration pairs at equivalence points

Type	pH	Example of salt formed
Strong acid–strong base	About 7	Sodium chloride
Strong acid–weak base	Lower than 7	Ammonium chloride
Weak acid–strong base	Higher than 7	Sodium ethanoate
Weak acid–weak base	About 7	Ammonium ethanoate

Titration calculations

Revised

Questions may be set that involve simple titration data.

Example 1

Involving an unknown concentration

$25.0\,cm^3$ of $0.100\,mol\,dm^{-3}$ hydrochloric acid are titrated against sodium hydroxide solution of an unknown concentration. It is found that $23.50\,cm^3$ of the alkali are required for equivalence. Calculate the concentration of the sodium hydroxide solution.

Answer

$$NaOH(aq) + HCl(aq) \rightarrow NaCl(aq) + H_2O(l)$$

$$\text{Moles of HCl(aq)} = \frac{\text{volume } (cm^3)}{1000\,cm^3} \times \text{concentration } (mol\,dm^{-3})$$

$$= \frac{25.0}{1000} \times 0.100$$

$= 2.50 \times 10^{-3}\,mol$

So the number of moles of NaOH must also be $2.50 \times 10^{-3}\,mol$ (1:1 ratio in the equation)

$$\text{Concentration of NaOH} = 2.50 \times 10^{-3} \times \frac{1000}{23.50}$$

$= 0.106\,mol\,dm^{-3}$

Example 2

Involving an unknown volume

25.0 cm^3 of 0.250 mol dm^{-3} sulfuric(VI) acid are titrated against potassium hydroxide solution of concentration 0.825 mol dm^{-3}. What volume of the potassium hydroxide solution is required for equivalence?

Answer

$$2KOH(aq) + H_2SO_4(aq) \rightarrow K_2SO_4(aq) + 2H_2O(l)$$

$$\text{Moles of } H_2SO_4(aq) = \frac{\text{volume (cm}^3)}{1000 \text{ cm}^3} \times \text{concentration (mol dm}^{-3})$$

$$= \frac{25.0}{1000} \times 0.250$$

$= 6.25 \times 10^{-3}$ mol

So the amount of KOH must be $2 \times 6.25 \times 10^{-3}$ mol = 0.0125 mol (2 : 1 ratio in the equation)

$$\text{Volume of KOH} = 0.0125 \times \frac{1000}{0.825}$$

$= 15.15$ cm^3

Indicators

Revised

Indicators are substances that are weak acids or weak bases and they are able to give a measure of the pH of a solution by their colour. There are many different indicators and they change from one colour to another at different pH values — these are called pK_{in} values.

Figure 3.2 shows the colours of some well-known indicators together with the pH range of values at which they change colour.

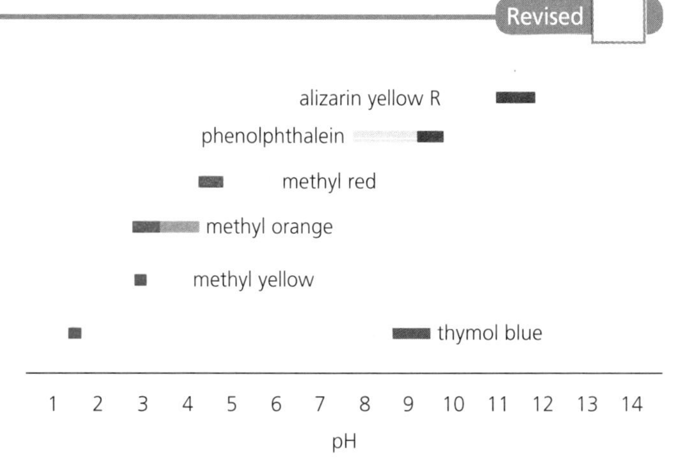

Figure 3.2 Some indicators and their colour change ranges

Choice of indicator

Revised

The choice of just which indicator to use in a particular titration depends on the pH at the equivalence point in the titration — the pH at the point at which the acid and base have reacted with each other exactly.

We choose an indicator that changes colour at a pH equal to, or very close to, the pH at the equivalence point of the acid–base reaction. This is always on the 'steep' part of the titration curve. This means that the colour change occurs with the addition of one drop of base and that the equivalence point is sharp.

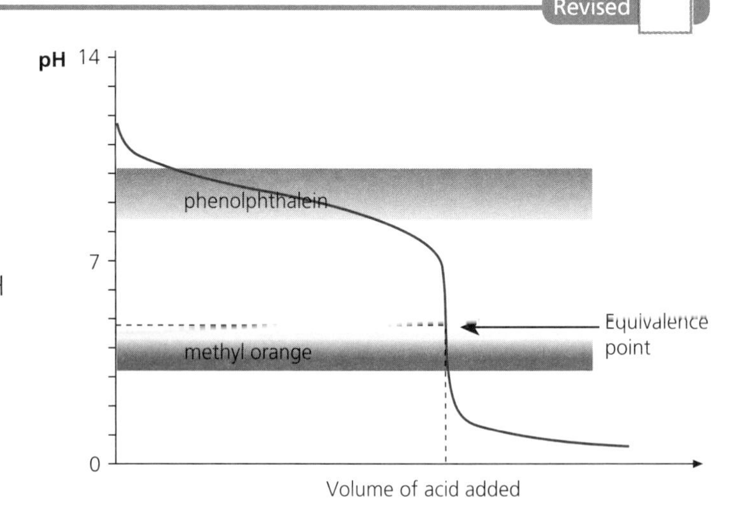

Figure 3.3 A typical titration curve for the neutralisation of a weak base by a strong acid

In the titration shown in Figure 3.3, between a strong acid and a weak base, the pH at equivalence is about 5. So we need to choose an indicator that changes colour at about this pH — from Figure 3.2 methyl orange looks a good choice. If we chose phenolphthalein, this has a pK_{in} value indicating a colour change at 8–10 and this would result in the indicator changing over the addition of a large volume of base with no sharp end-point.

Examiners' tip

Choose an indicator so that its pK_{in} value is never more than 1 pH unit from the equivalence pH. So, if the pH at equivalence is 9.20, the pK_{in} of the indicator should be between 8.20 and 10.20.

Buffer solutions

Buffer solutions exist as one of two main types according to their components.

- **Acid buffer:** a solution containing a mixture of a weak acid and its conjugate base. The pH of the buffer solution is lower than 7.
  Example: ethanoic acid, $CH_3COOH(aq)$, and sodium ethanoate, $CH_3COO^-Na^+$.
- **Basic buffer:** a solution containing a mixture of a weak base and its conjugate acid. The pH of the buffer solution is higher than 7.
  Example: ammonia solution, $NH_3(aq)$, and ammonium chloride, NH_4Cl.

Buffer solutions are defined as solutions that resist a change in pH on adding small amounts of either acid or base (or by diluting the solution by adding water).

The pH at which a buffer operates most effectively depends on the pK_a of the acidic component present.

For a buffer to operate effectively, it is crucial that the mixture contains large amounts of both the base and the acid.

Examiners' tip

Remember that any carboxylic acid and one of its corresponding salts, can be used (when combined) to make a buffer.

How does a buffer work?

Revised

Consider the effect of adding acid and base to an ethanoic acid–sodium ethanoate buffer system (an acid buffer). The equilibrium operating is:

$$CH_3COOH(aq) + H_2O(l) \rightleftharpoons CH_3COO^-(aq) + H_3O^+(aq)$$

On adding a base

Added hydroxide ions react with the hydrogen ions, $H_3O^+(aq)$, in the above equilibrium:

$$H_3O^+(aq) + OH^-(aq) \rightarrow 2H_2O(l)$$

This reduces the hydrogen ion concentration in the equilibrium mixture. The equilibrium consequently shifts to the right-hand side by dissociating some of its molecular ethanoic acid to produce more hydrogen ions:

$$CH_3COOH(aq) + H_2O(l) \rightarrow CH_3COO^-(aq) + H_3O^+(aq)$$

The final pH in the new equilibrium mixture is similar to its starting pH and the value for K_a in the equilibrium mixture is preserved (remember that changes in concentration do not affect K_a or any other equilibrium constant).

Examiners' tip

Always write an equation for the dissociation of the *acid* — a carboxylic acid or ammonium ion — and then use that equilibrium to explain how the buffer works.

On adding an acid

$$CH_3COOH(aq) + H_2O(l) \rightleftharpoons CH_3COO^-(aq) + H_3O^+(aq)$$

Adding an acid increases the concentration of the hydrogen ions in the equilibrium above. The equilibrium shifts to the left-hand side using up ethanoate ions, CH_3COO^-, to remove the excess hydrogen ions:

$$CH_3COO^-(aq) + H_3O^+(aq) \rightarrow CH_3COOH(aq) + H_2O(l)$$

The concentration of hydrogen ions in the new equilibrium mixture will be similar to that at the start and the pH remains about the same.

Another commonly used buffer is the ammonia–ammonium ion system and the equilibrium operating is:

$$NH_4^+(aq) + H_2O(l) \rightleftharpoons NH_3(aq) + H_3O^+(aq)$$

On adding acid, the reaction shifts to the left-hand side; on adding alkali, the reaction shifts to the right-hand side. The explanation is the same as for the ethanoic acid–sodium ethanoate system.

Calculations involving buffer solutions

Revised

Consider the general dissociation of a weak acid, HA:

$$HA(aq) \rightleftharpoons A^-(aq) + H^+(aq)$$

$$K_a = \frac{[H^+(aq)][A^-(aq)]}{[HA(aq)]}$$

A question may require you to calculate the pH of a buffer solution in which the concentration of both components are given, that is [A⁻] and [HA], together with the pK_a for the acid present.

Example 1

Calculate the pH of a buffer solution that contains a mixture of 0.0500 mol dm^{-3} HCOOH(aq) and 0.0800 mol dm^{-3} HCOONa(aq). The pK_a of methanoic acid is 3.75.

Answer

$$HCOOH(l) \rightleftharpoons HCOO^-(aq) + H^+(aq)$$

$$K_a = \frac{[HCOO^-(aq)][H^+(aq)]}{[HCOOH(aq)]}$$

pK_a for the acid is 3.75, so:

$K_a = 10^{-3.75}$

$= 1.78 \times 10^{-4}$ mol dm^{-3}

Substituting into the expression for K_a gives:

$$1.78 \times 10^{-4} = 0.0800 \times \frac{[H^+(aq)]}{0.0500}$$

$$[H^+(aq)] = 1.78 \times 10^{-4} \times \frac{0.0500}{0.0800}$$

$= 1.11 \times 10^{-4}$ mol dm^{-3}

$pH = -\log_{10}[H^+(aq)]$

$= -\log_{10}(1.11 \times 10^{-4})$

$= 3.95$

Exam practice

1 This question is about the pH of some solutions containing potassium hydroxide or methanoic acid or both. Give all pH values to 2 decimal places.

(a) (i) Write an expression for pH. [1]

(ii) Write an expression for the ionic product of water. [1]

(iii) At 10°C, a 0.132 mol dm^{-3} solution of potassium hydroxide has a pH of 13.22. Calculate the value of K_w at 10°C. [2]

(b) (i) Write an expression for K_a for methanoic acid, HCOOH(aq). [1]

(ii) Calculate the pH of a solution of methanoic acid of concentration 0.200 mol dm^{-3}. pK_a for methanoic acid is 3.75. [3]

(c) (i) What is meant by a 'buffer solution'? [1]

(ii) Calculate the pH of the solution formed when 20.0 cm^3 of 0.200 mol dm^{-3} methanoic acid is added to 10.0 cm^3 of 0.132 mol dm^{-3} potassium hydroxide solution. [4]

2 A titration is carried out in which 25.0 cm^3 of benzoic acid of concentration 0.100 mol dm^{-3} is titrated against sodium hydroxide solution of an unknown concentration. It is found that 15.50 cm^3 of sodium hydroxide is required for complete reaction. (pK_a for benzoic acid is 4.20)

(a) (i) Write an equation for the reaction taking place. [1]

(ii) Calculate the concentration of the sodium hydroxide solution used. [2]

(iii) Sketch a titration curve for the reaction taking place. Label your axes carefully indicating the pH at which equivalence takes place. [2]

(iv) Give the name of an indicator that could be used in this titration. Select from the list in Figure 3.2 and explain your choice. [2]

(b) In another titration, 25.0 cm^3 of ethanoic acid (pK_a = 4.76) of concentration 0.100 mol dm^{-3} was used instead of 25.0 cm^3 of benzoic acid. State how the volume of sodium hydroxide solution (of the same concentration) required for neutralisation would change, if at all. [2]

Answers and quick quizzes online

Online

Examiners' summary

You should now have an understanding of:

- ✔ Brønsted–Lowry acid and bases
- ✔ pH
- ✔ how to calculate the pH of solutions of strong acids and strong bases
- ✔ the ionic product of water, K_w
- ✔ weak acids and weak bases
- ✔ acid dissociation constant, K_a
- ✔ how to calculate the pH of a solution of a weak acid
- ✔ pH titration curves
- ✔ indicators
- ✔ buffers and how to calculate the pH of an acid buffer solution

4 Isomerism in organic chemistry

Structural isomers

Structural isomers are defined as compounds that have the same molecular formula but different structures.

For example, the structural isomers of C_5H_{12} with different carbon chains are shown in Figure 4.1. They are called pentane, 2-methylbutane and 2,2-dimethylpropane respectively.

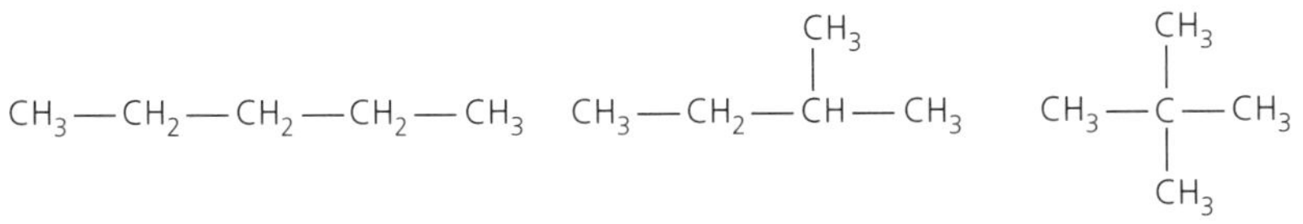

$CH_3—CH_2—CH_2—CH_2—CH_3$

pentane

$CH_3—CH_2—CH(CH_3)—CH_3$

2–methylbutane

$CH_3—C(CH_3)_2—CH_3$

2,2–dimethylpropane

Figure 4.1 Isomers of C_5H_{12}

Figure 4.2 shows two molecules that are also structural isomers, but these have different functional groups. The first is an alcohol and the second is an ether. Notice how both molecules consist of atoms some of which are bonded to **different 'neighbouring' atoms**.

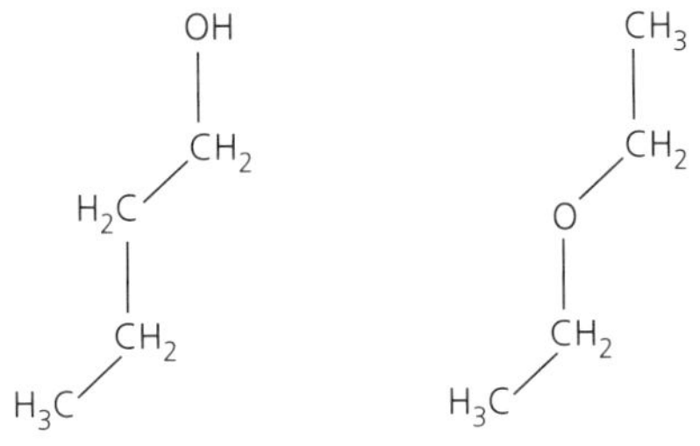

$H_3C—CH_2—CH_2—CH_2—OH$

butan-1-ol

$H_3C—CH_2—O—CH_2—CH_3$

ethoxyethane

Figure 4.2 Functional group isomers of $C_4H_{10}O$

Structural isomers have different physical and chemical properties because their structures are different and so their intermolecular forces differ in type and in strength.

Stereoisomerism

Stereoisomers are compounds that have the same structural formula, but their atoms differ in **spatial arrangement**.

- Atoms in molecules of this type are bonded to the **same 'neighbouring' atoms**, but their 3-dimensional arrangement is different.
- There are two types of stereoisomerism — **E–Z isomerism** and **optical isomerism**.

E–Z isomerism

Revised

This type of isomerism exists due to the restricted rotation about a carbon–carbon double bond. The π bonding electrons prevent free rotation, and this means that two groups, or substituents, may either be on the same side of the double bond (Z-isomer) or on opposite sides (E-isomer).

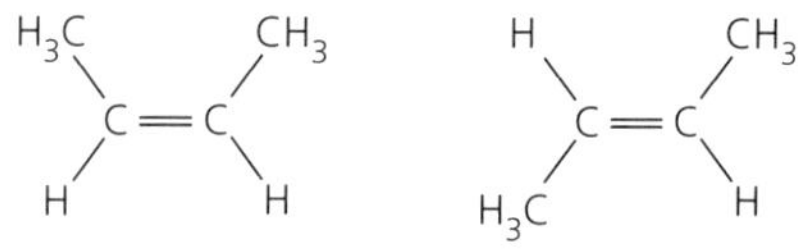

For example, Figure 4.3 shows the isomerism of but-2-ene in which the methyl groups either side of the carbon–carbon double bond are on opposite sides (E) or on the same side (Z).

Z-but-2-ene E-but-2-ene

Figure 4.3 E–Z isomers of but-2-ene

Figure 4.4 shows another example of E-Z isomerism using 1,2-dichloroethene.

Z-1,2-dichloroethene E-1,2-dichloroethene

Figure 4.4 E–Z isomers of 1,2-dichloroethene

The boiling points of the dichloro-isomers are 60.3°C and 47.5°C respectively. In the case of Z-1,2-dichloroethene, the $^{\delta+}C–Cl^{\delta-}$ dipoles do not cancel out so this molecule has an overall dipole. In the case of E-1,2-dichloroethene, the $^{\delta+}C–Cl^{\delta-}$ dipoles do cancel and so this molecule is non-polar. The intermolecular forces between Z-1,2-dichloroethene (dipole–dipole forces) molecules will be stronger than the van der Waals' forces between molecules of E-1,2-dichloroethene and the boiling point of the E-isomer will be lower.

Optical isomerism

Revised

Because molecules are 3-dimensional, many are different compared to their mirror image molecules. Molecules like this that have non-superimposable mirror images are called **chiral molecules** or **enantiomers**.

One of the enantiomers will **rotate the plane of polarised light** to the right, and the other will rotate it to the left. The signs (+) and (–) are used to show that a molecule rotates plane-polarised light to the right or the left respectively.

When carbon atoms forms bonds to four different groups arranged as a tetrahedron in space (or have an **asymmetric carbon atom**), enantiomers will form.

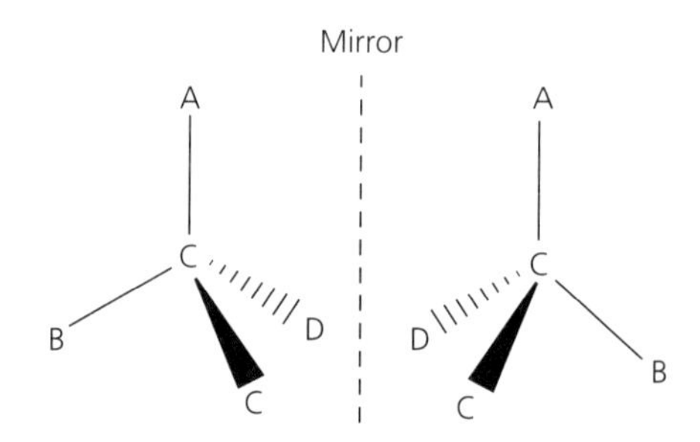

Figure 4.5 Optical isomers

The molecules in Figure 4.5 are mirror images of each other, but they are non-superimposable. These are therefore different molecules and are called optical isomers.

Figure 4.6 shows the amino acid called alanine (2-aminopropanoic acid) with the structural formula $H_2NCH(CH_3)COOH$. Again, the mirror image molecules cannot be superimposed on each other so alanine exists as two enantiomers.

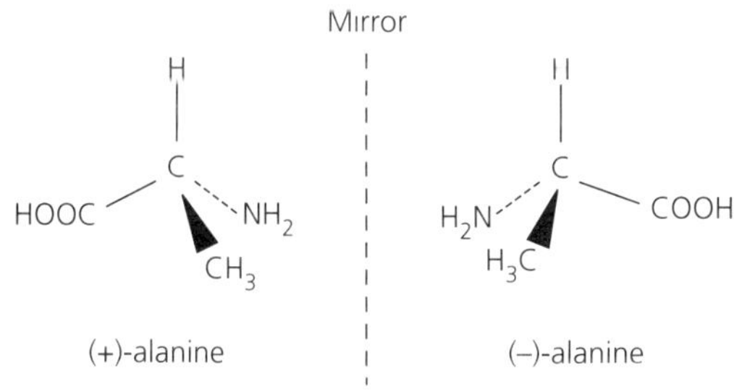

Figure 4.6 Optical isomerism in amino acids

Examiners' tip

When drawing the structures of enantiomers, make sure that you use 3-dimensional diagrams.

Enantiomers in reactions

Revised

Many pharmaceutical drugs consist of molecules that are chiral, and their chirality is an essential feature of their chemical behaviour.

Often one of the enantiomeric forms has a markedly different chemical behaviour from the other form. This can be illustrated using the example of thalidomide — a drug used in the 1960s to cure nausea in pregnancy. The structures of the mirror image molecules are shown in Figure 4.7.

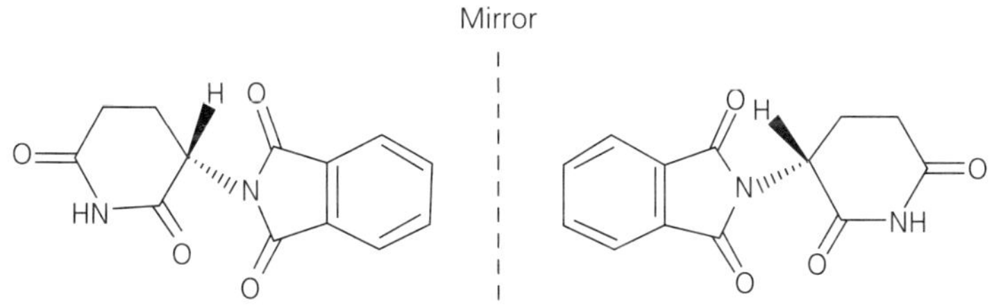

Figure 4.7 Thalidomide

The molecule on the left reduces morning sickness — the desired effect — and the molecule on the right results in deformation of the foetus. Although the molecules look similar, subtle differences in their stereochemistry have a huge difference on their chemical behaviour.

In this case, it is necessary to separate one enantiomer from the other before distribution for use by pregnant women. This can be expensive because it often requires the use of an additional enzymic process to remove the unwanted molecules from the reaction mixture.

Alternatively, the synthesis of a molecule can have its stereochemistry controlled so that only one of the enantiomers is formed. This technique often uses chiral reagents and can be time consuming.

In a reaction that does not have its stereochemistry controlled, it is common to produce a mixture of both the (+) and (–) optical isomers — this is called a racemic mixture or racemate. The mixture does not affect plane-polarised light because each enantiomer has an equal and opposite effect on the optical rotation of the plane-polarised light.

In many reaction mechanisms, it may be equally possible for an attacking species to react on either side of an intermediate — a species that is temporarily formed during a reaction and has a short half-life. As the reaction then takes place, it may be possible to form an equal mixture of both the (+) and (–) forms.

Now test yourself

Tested

1 **(a)** What is meant by 'structural isomers'?
(b) Draw the structural formulae and name the structural isomers of C_6H_{14}.

2 **(a)** Write the molecular formula of this fatty acid:

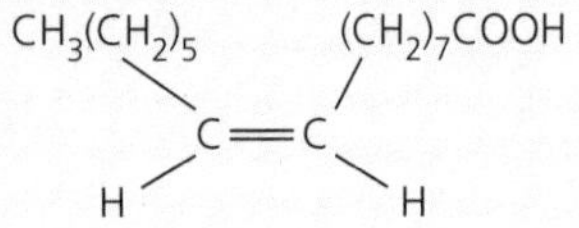

$CH_3(CH_2)_5$ (upper left), $(CH_2)_7COOH$ (upper right), C=C, H (lower left), H (lower right)

(b) The molecule in part (a) can show E–Z isomerism. Explain why.
(c) The isomer shown in the diagram is the Z-isomer. Draw the structure of the E-isomer.

3 **(a)** What is meant by the term 'stereoisomers'?
(b) Lactic acid has the structural formula $HOOCCH(OH)CH_3$. Draw the enantiomeric forms of lactic acid.
(c) Describe a technique by which the two enantiomeric forms of lactic acid can be distinguished.

Answers on p. 108

Exam practice

1 **(a)** Name molecule, A. [1]

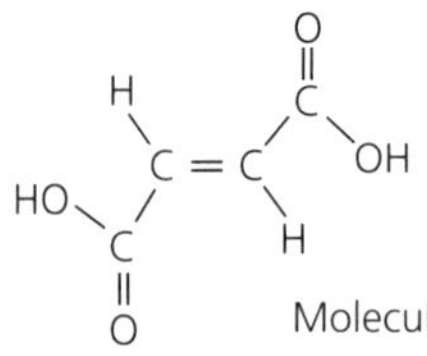

H and HOOC on one carbon, COOH and H on the other, joined by C=C (H and COOH on the upper side, HOOC and H on the lower side)

Molecule A

(b) What type of stereoisomerism can molecule A display? Explain your answer. [2]
(c) Draw the displayed formula of the other stereoisomer of A. [1]
(d) These stereoisomers have different melting points — state which has the higher melting point. Explain your answer. [2]

Answers and quick quizzes online

Online

Examiners' summary

You should now have an understanding of:

- ✔ structural isomers
- ✔ E–Z isomerism
- ✔ optical isomerism
- ✔ the meaning of the terms 'enantiomer' and 'racemate'
- ✔ how to draw various isomers
- ✔ why the formation and use of specific enantiomers is important, especially in medicine

5 Compounds containing the carbonyl group

Aldehydes and ketones

Aldehydes and ketones are described as **carbonyl compounds** because they contain the carbonyl group (Figure 5.1).

Figure 5.1 The carbonyl group

Figure 5.2 shows the functional groups in an aldehyde and a ketone.

The molecule in Figure 5.3 is called ethanal — it is an aldehyde because it contains the –CHO functional group.

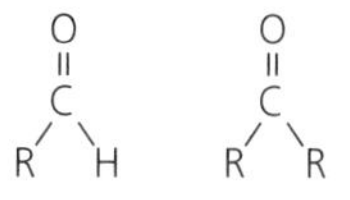

aldehyde ketone

Figure 5.2 Two types of carbonyl functional groups

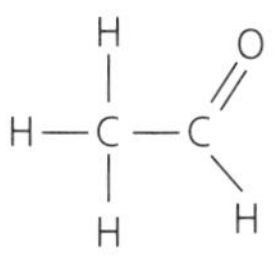

Figure 5.3 Ethanal

Other more complex naturally occurring aldehydes and ketones are shown in Figure 5.4.

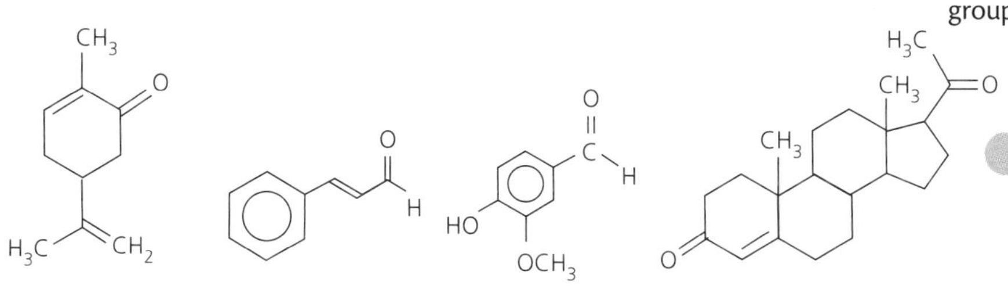

carvone (spearmint and caraway)
cinnamaldehyde (cinnamon bark)
vanilin (vanilla bean)
progesterone (female sex hormone)

Figure 5.4 Some naturally occurring carbonyl compounds (drawn as skeletal diagrams)

Now test yourself

1 Which of the molecules in Figure 5.4 are aldehydes and which are ketones?

Answers on p. 109

Tested

Reactions of aldehydes and ketones

Revised

Redox processes

At AS level, one method for synthesising aldehydes and ketones involves oxidising alcohols using acidified potassium dichromate(VI). Primary alcohols are oxidised to form aldehydes and then carboxylic acids. Secondary alcohols are oxidised to form ketones. For example, using propan-1-ol, $CH_3CH_2CH_2OH$:

$$\underset{\text{propan-1-ol}}{CH_3CH_2CH_2OH} + [O] \rightarrow \underset{\text{propanal}}{CH_3CH_2CHO} + H_2O$$

Then, if sufficient oxidising agent [O] is present:

$$\underset{\text{propanal}}{CH_3CH_2CHO} + [O] \rightarrow \underset{\text{propanoic acid}}{CH_3CH_2COOH}$$

Aldehydes are easily oxidised to form the corresponding carboxylic acid. This means that aldehydes are very good **reducing agents** and will, therefore, reduce other species, for example Ag^+ to Ag, or Cu(II) compounds to Cu(I) compounds.

Fehling's solution

Fehling's solution contains copper(II) ions and these are reduced to form copper(I) compounds such as copper(I) oxide. When an aldehyde is added to Fehling's solution and the mixture is warmed, a **red precipitate** is formed:

$$2Cu^{2+}(aq) + H_2O(l) + 2e^- \rightarrow Cu_2O(s) + 2H^+(aq)$$

The aldehyde is oxidised to form a carboxylic acid, for example:

$$CH_3CHO(aq) + [O] \rightarrow CH_3COOH(aq)$$

$$CH_3CHO(aq) + H_2O(l) \rightarrow CH_3COOH(aq) + 2H^+(aq) + 2e^-$$

Fehling's solution and **Tollens' reagent** can be used to distinguish between an aldehyde and a ketone. Ketones do not react with these reagents.

Tollens' reagent

Tollens' reagent contains the complex ion $[Ag(NH_3)_2]^+$. It is formed by adding ammonia solution dropwise to silver nitrate solution until the brown precipitate of silver oxide just dissolves to form a colourless solution.

When this is warmed with an aldehyde, a **silver mirror** is formed. The silver ions are reduced in the reaction:

$$Ag^+(aq) + e^- \rightarrow Ag(s)$$

The aldehyde is oxidised in the reaction. Ketones are not easily oxidised and so do not react.

Nucleophilic addition reactions

The carbonyl group, C=O, is unsaturated so addition reactions can happen. The carbon atom of the group is less electronegative than the oxygen atom so the bond is polarised and the carbon atom has a $\delta+$ charge (Figure 5.5) so **nucleophiles** attack it. Hence, carbonyl compounds can take part in nucleophilic addition reactions.

A **nucleophile** seeks out centres of positive charge and is a lone-pair donor.

Figure 5.5 The polar carbonyl group

The nucleophile performs an addition process on the carbonyl group. A general type of nucleophilic addition is shown in Figure 5.6.

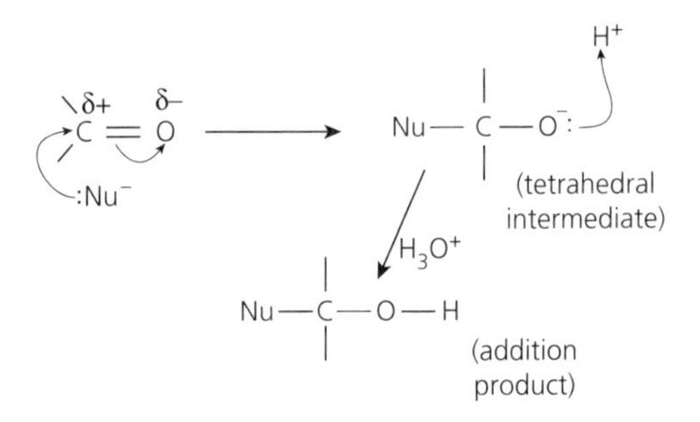

Figure 5.6 Nucleophilic addition mechanism

Reduction with sodium tetrahydridoborate(III), $NaBH_4$

$NaBH_4$ in aqueous solution behaves as a reducing agent. It provides hydride ions, H^-, as nucleophiles and these react with carbonyl groups in both aldehydes and ketones to form the corresponding primary and secondary alcohols.

The mechanism is a nucleophilic addition process — see Figure 5.7:

- The hydride ion, H^-, donates its lone pair to the slightly positively charged carbon atom in the carbonyl group.
- The higher energy π bonding electrons in the carbonyl group move onto the oxygen atom to form a negative charge.
- One of the lone pairs of electrons on the oxygen atom is used to allow protonation, normally by an acid. This forms a hydroxyl group, OH.

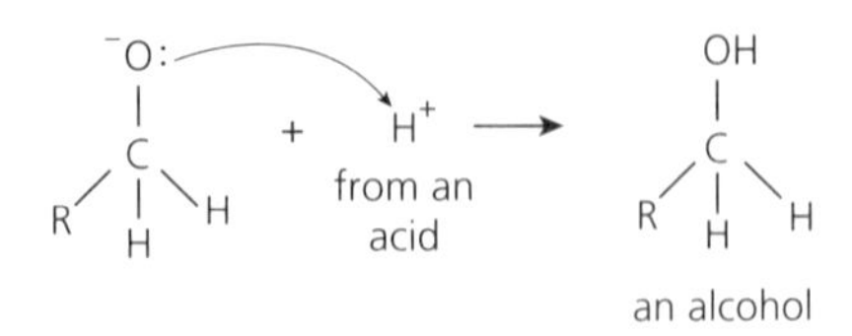

Figure 5.7 Nucleophilic addition mechanism reducing an aldehyde

Reaction with hydrogen cyanide, HCN

It is possible to use the cyanide ion, CN^-, as a nucleophile when reacting with a carbonyl group. It reacts in the same way as the hydride ion, H^-.

The cyanide ion, CN^-, attacks the carbonyl carbon atom in this process to form a new substance called a **hydroxynitrile**.

Using **hydrogen cyanide** and cyanides (in general) is highly hazardous due to the highly toxic nature of these substances.

Using ethanal as an example:

$$CH_3CHO + HCN \rightarrow CH_3CH(OH)(CN)$$

HCN is a very weak acid and so provides few CN^- ions — a trace of soluble KCN provides the nucleophile.

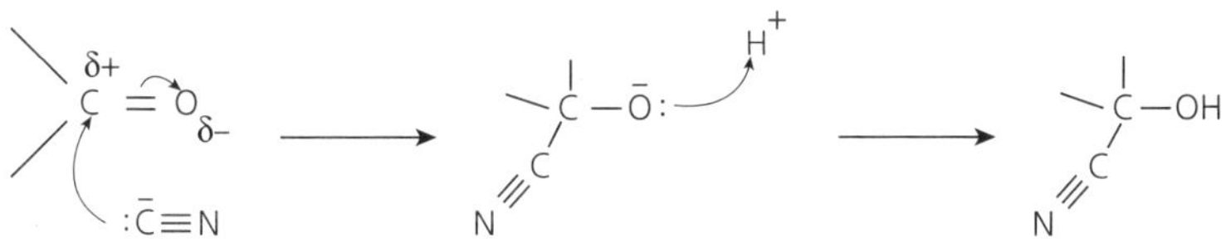

Figure 5.8 Nucleophilic addition mechanism forming a hydroxynitrile

In the first stage of this mechanism (Figure 5.8), the cyanide ion can approach the trigonal planar carbonyl group from either side. This means that the stereochemistry is not controlled and the resulting hydroxynitrile is likely to form as a racemic mixture in which a 50:50 mixture of enantiomers forms.

Now test yourself Tested

3 Draw the structure of the product formed when the following reacts with hydrogen cyanide (HCN) in KCN.

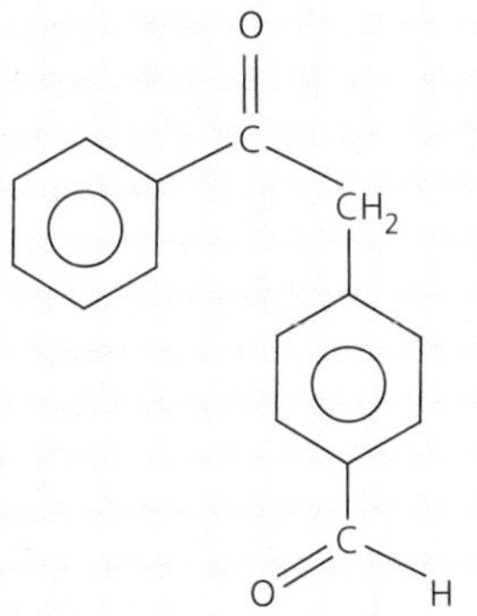

Answers on p. 109

Now test yourself

2 Draw the structures of the products formed when the following are reduced using $NaBH_4$.

(a)

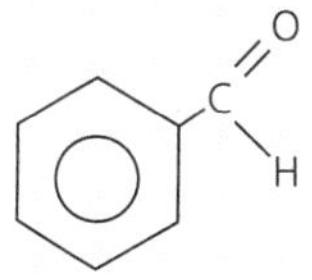

(b)

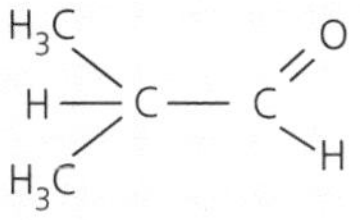

(c) $CH_3—CH_2—CH_2—CHO$

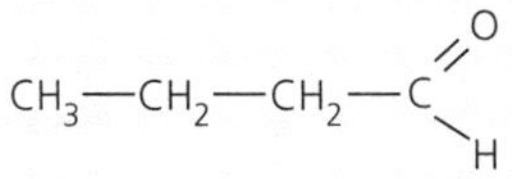

Answers on p. 109

Tested

Carboxylic acids and esters

Carboxylic acids are molecules containing the functional group shown in Figure 5.9.

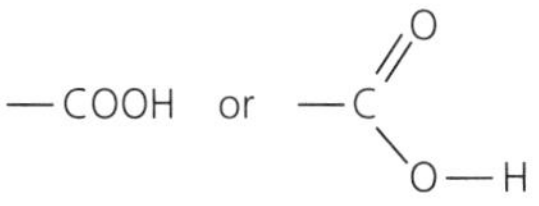

—COOH or —C(=O)—O—H

Figure 5.9 Carboxylic acid functional group

The carbonyl group, C=O, withdraws electrons from the hydroxyl group making the O–H bond weaker. This induces dissociation to form the **carboxylate anion** and a hydrated proton — for example:

$CH_3COOH(aq) + H_2O(l) \rightleftharpoons CH_3COO^-(aq) + H_3O^+(aq)$

Carboxylic acids are **weak acids** because they dissociate only partially in solution.

Chemical reactions of carboxylic acids

Revised

Acidic behaviour

Carboxylic acids show all the normal reactions of acids.

- With **reactive metals** they form a carboxylate salt and hydrogen:
  $Na(s) + CH_3COOH(aq) \rightarrow CH_3COO^-Na^+(aq) + \frac{1}{2}H_2(g)$
  or $Na(s) + H^+(aq) \rightarrow Na^+(aq) + \frac{1}{2}H_2(g)$
- With **metal oxides** they form a salt and water:
  $CuO(s) + 2CH_3COOH(aq) \rightarrow (CH_3COO)_2Cu(aq) + H_2O(l)$

Carboxylic acids react with metal carbonates to form carbon dioxide gas. This reaction can be used to test for the presence of a carboxylic acid.

- With **alkalis** they form a salt and water:
  $NaOH(aq) + CH_3COOH(aq) \rightarrow CH_3COO^-Na^+(aq) + H_2O(l)$
- with **metal carbonates** they form a salt, water and carbon dioxide:
  $K_2CO_3(s) + 2CH_3COOH(aq) \rightarrow 2CH_3COO^-K^+(aq) + H_2O(l) + CO_2(g)$

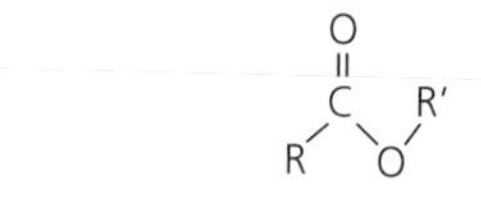

Figure 5.10 The ester functional group

Reaction of carboxylic acids with alcohols to make esters

Carboxylic acids with react with alcohols, in the presence of **concentrated sulfuric(VI) acid** as catalyst, to form an equilibrium mixture in which the products are an ester and water.

$$HCOOH + H_3C{-}CH_2OH \rightleftharpoons HCOOCH_2CH_3 + H_2O$$

methanoic acid | ethanol | ethyl methanoate

Figure 5.11 An esterification reaction

Esters contain the functional group shown in Figure 5.10.

In the example shown in Figure 5.11, methanoic acid reacts with ethanol in the presence of concentrated sulfuric(VI) acid to form an ester called ethyl methanoate and water.

In the esterification process, water is formed by combination of the H from the hydroxyl group of the alcohol and the O–H group from the carboxylic acid, as illustrated by the reaction of butanoic acid and methanol in Figure 5.12.

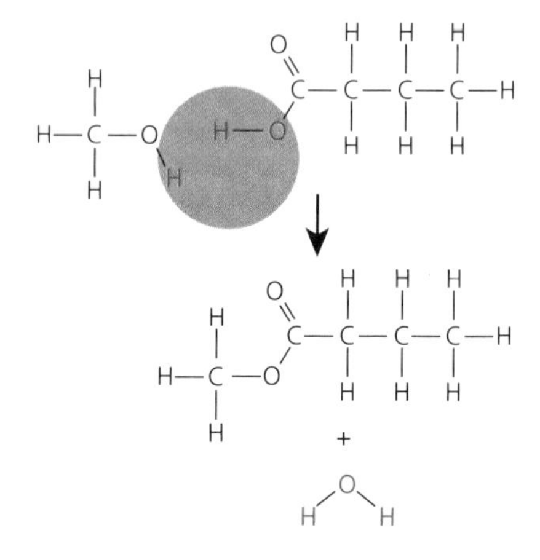

Figure 5.12 The esterification process

Esters are used in food and in the perfume industry as well as being good solvents and plasticisers.

Reactions of esters

Revised

Hydrolysis reactions involving esters can take place in which either acids or bases may be used to catalyse the reactions. In these reactions, the acid or the base is a catalyst so the processes are called **acid- or base-catalysed hydrolysis** reactions.

It is important to note that:

- in basic conditions, esters undergo complete hydrolysis forming the corresponding alcohol and the sodium salt of the carboxylic acid
- in acidic conditions, esters are not completely hydrolysed — an equilibrium mixture is formed in which some ester is still present.

Examiners' tip

When esters are hydrolysed in acid or basic solution, an alcohol is always formed but the carboxylic acid only forms in acidic conditions. A carboxylate anion forms in basic conditions.

Base-catalysed hydrolysis

Figure 5.13 shows the hydrolysis of methyl ethanoate in basic conditions as an example.

$$H_3C{-}O{-}C(=O){-}CH_3 + OH^- \rightarrow CH_3OH + {}^-O{-}C(=O){-}CH_3$$

methyl ethanoate | methanol | ethanoate ion

Figure 5.13 Base hydrolysis of an ester

The carboxylate anion, CH_3COO^-, can now react with water to reform the hydroxide ion catalyst, as well as ethanoic acid.

Acid-catalysed hydrolysis

Figure 5.14 shows the hydrolysis of methyl ethanoate in acidic conditions as an example.

$$H_3C{-}O{-}C({=}O){-}CH_3 + H_2O \overset{H^+}{\rightleftharpoons} CH_3OH + HO{-}C({=}O){-}CH_3$$

methyl ethanoate — methanol — ethanoic acid

Figure 5.14 Acid hydrolysis of an ester

Naturally occurring fats and **oils** are esters made from propane-1,2,3-triol (an alcohol commonly known as glycerol) and fatty acids (long-chain saturated or unsaturated carboxylic acids). If fats or oils are heated in concentrated **sodium hydroxide** solution (Figure 5.15), the ester is **hydrolysed** to form propane-1,2,3-triol (which is used in the food and cosmetics industries) and the sodium salts of the fatty acids (which are used in making soap).

$$\begin{matrix} CH_2O{-}C({=}O){-}R \\ | \\ CHO{-}C({=}O){-}R' \\ | \\ CH_2O{-}C({=}O){-}R'' \end{matrix} + 3NaOH \longrightarrow \begin{matrix} CH_2OH \\ | \\ CHOH \\ | \\ CH_2OH \end{matrix} + \begin{matrix} Na^+\ {}^-O{-}C({=}O){-}R \\ Na^+\ {}^-O{-}C({=}O){-}R' \\ Na^+\ {}^-O{-}C({=}O){-}R'' \end{matrix} + 3H_2O$$

glycerol — sodium salts of long-chain fatty acids

Figure 5.15 Soap making

> **Now test yourself**
>
> 4 Draw the structure of the products expected when the ester shown below is hydrolysed with (a) sodium hydroxide solution and (b) sulfuric(VI) acid.
>
> $CH_3CH_2{-}C({=}O){-}O{-}CH_2CH_3$
>
> Answers on p. 109
>
> Tested

Biodiesel

Revised

When a naturally occurring fat or oil is heated with methanol (Figure 5.16) in the presence of a catalytic quantity of acid, glycerol (propane-1,2,3-triol) and methyl esters are formed. The methyl esters can then be separated from the mixture and added to normal diesel to make **biodiesel** or used 'pure' as a fuel.

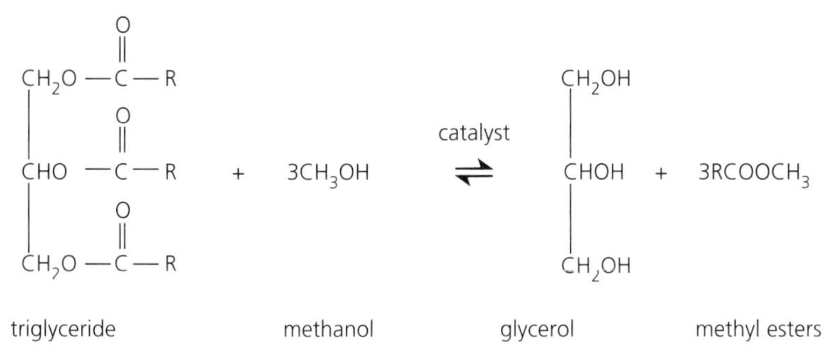

Figure 5.16 A source of biodiesel

> **Revision activity**
>
> Test yourself on organic reactions by preparing sheets with 3 columns — the 'starting molecule', the 'reagent and conditions' and the 'product molecule'. Then cover up one of the columns and try to work it out from the remaining information in the other two columns. Do this every time you meet a new group of organic compounds.

Acylation

Acylation is the process of replacing a hydrogen atom in certain molecules by an RCO group, where R is an alkyl group. The structure of an acyl group is shown in Figure 5.17.

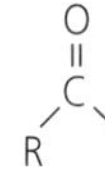

Figure 5.17 The acyl functional group

Acylations can be carried out using acyl chlorides such as ethanoyl chloride, CH_3COCl, or by acid anhydrides such as ethanoic anhydride, $(CH_3CO)_2O$. Figure 5.18 shows the structure of these two reagents.

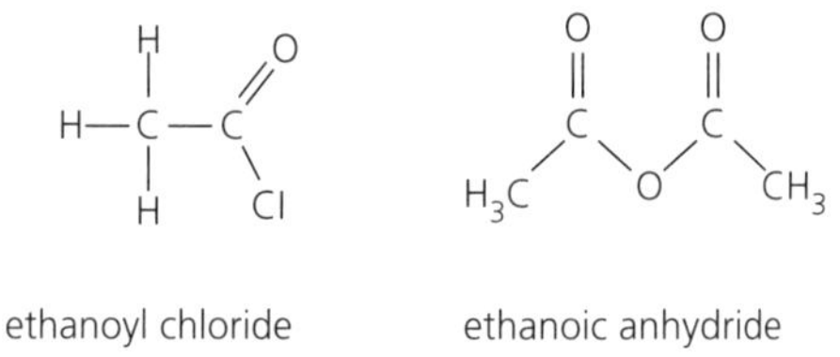

ethanoyl chloride   ethanoic anhydride

Figure 5.18 Acylating agents

Reactions of ethanoyl chloride and ethanoic anhydride

Revised

Nucleophiles such as water (H_2O), alcohols (ROH), ammonia (NH_3) and amines (RNH_2) all react with acyl chlorides and acid anhydrides in a predictable way.

Figure 5.19 shows the mechanism of the reaction between a nucleophile and an acyl chloride.

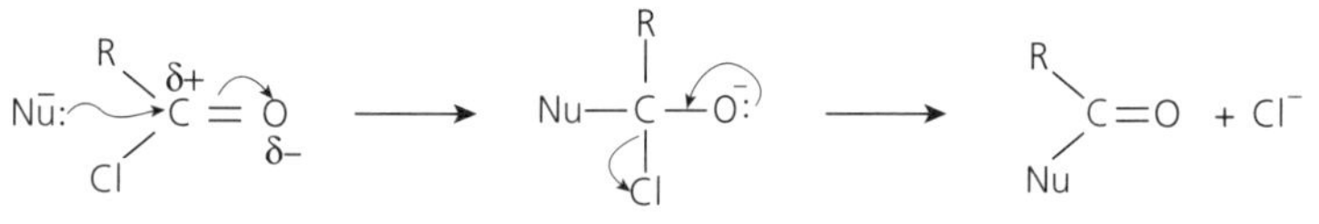

intermediate

Figure 5.19 The mechanism of nucleophilic addition–elimination

This mechanism is called **nucleophilic addition–elimination**. The successive stages are:

- The nucleophile donates a lone pair of electrons making a bond with the electron-deficient carbon atom of the carbonyl group.
- The high-energy π electrons in the C=O bond move to the oxygen atom to generate an intermediate anion.
- A lone pair belonging to the oxygen atom then moves back into the C–O bond to regenerate the C=O bond.
- The C–Cl bond breaks to release a chloride ion, Cl^-.

The overall effect is that the Cl group is replaced by a nucleophile.

Revision activity

Learn these mechanisms and try to understand what is happening in each stage. In examinations, you may be asked to work through unfamiliar mechanisms in which you are provided with a general scheme. On separate cards, draw a summary of each reaction mechanism you meet so that you'll have a complete set at the end.

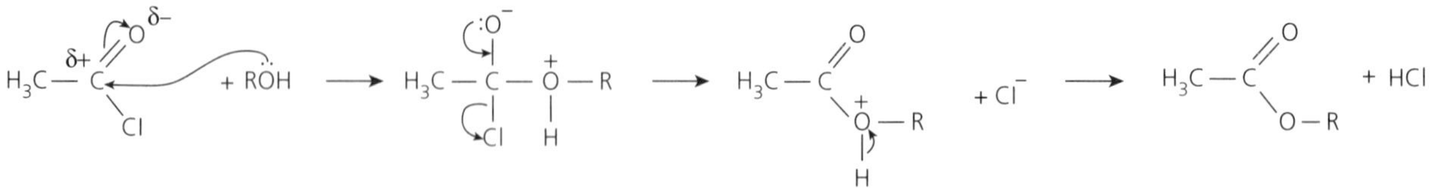

When ethanoyl chloride reacts with an alcohol, the mechanism shown in Figure 5.20 occurs. The lone pair on the oxygen atom of the alcohol is used to bond the nucleophile to the carbon atom of the carbonyl group.

Figure 5.20 The acylation of an alcohol

When salicylic acid reacts with ethanoic anhydride (Figure 5.21), the oxygen atom of the hydroxyl group in salicylic acid donates a lone pair to one of the carbon atoms in a carbonyl group of the ethanoic anhydride. This triggers a rupturing of the ethanoic anhydride molecule so that the salicylate group can replace the ethanoate group.

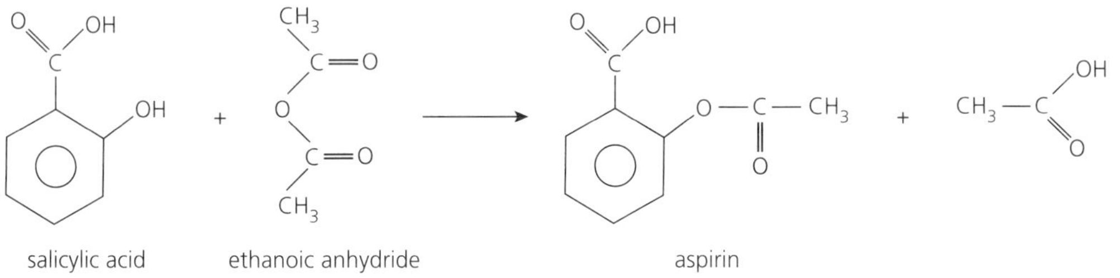

Figure 5.21 The synthesis of aspirin

When synthesising aspirin industrially, ethanoic anhydride is used instead of ethanoyl chloride because its reactions are more easily controlled. Ethanoyl chloride is difficult to store because it is easily hydrolysed by moisture and it releases corrosive HCl fumes.

Reactions of ethanoyl chloride

Ethanoyl chloride reacts with water, alcohols, ammonia and amines as follows.

- With water, H_2O
  $CH_3COCl + H_2O \rightarrow CH_3COOH + HCl$
  making ethanoic acid — a carboxylic acid.
- With methanol, CH_3OH
  $CH_3COCl + CH_3OH \rightarrow CH_3COOCH_3 + HCl$
  making methyl ethanoate — an ester.
- With ammonia, NH_3
  $CH_3COCl + 2NH_3 \rightarrow CH_3CONH_2 + NH_4Cl$
  making ethanamide — an amide.
- With methylamine, CH_3NH_2
  $CH_3COCl + CH_3NH_2 \rightarrow CH_3CONHCH_3 + HCl$
  making an N-substituted amide.

This is shown schematically in Figure 5.22.

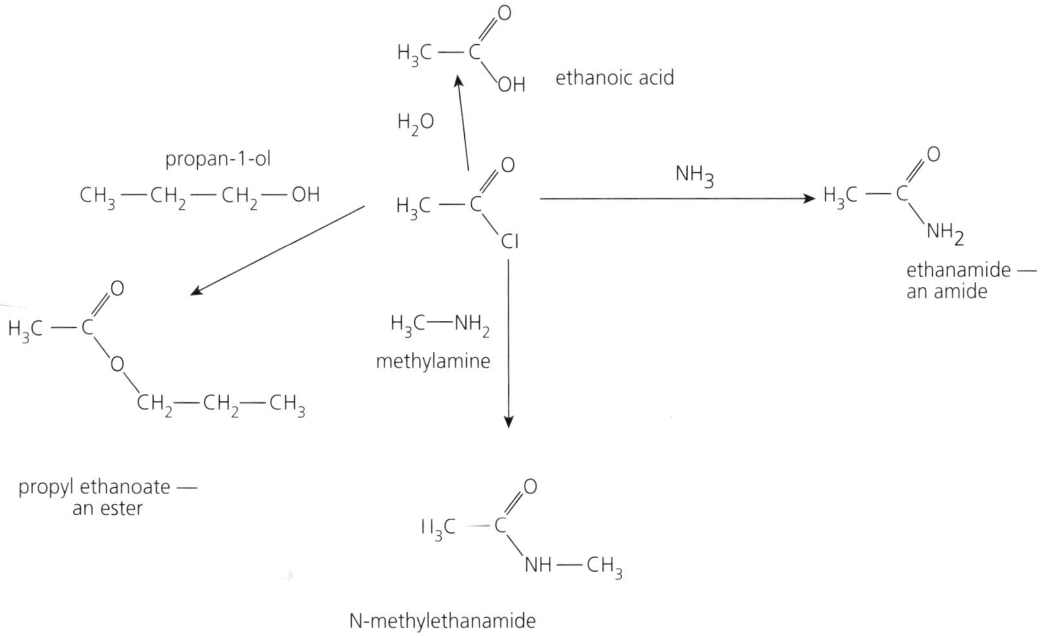

Figure 5.22 Reactions of ethanoyl chloride

Reactions of ethanoic anhydride

Ethanoic anhydride reacts with water, alcohols, ammonia and amines as follows.

Notice how the main organic product is the same as that formed by ethanoyl chloride, the only difference is that CH_3COOH is formed rather than HCl.

- With water, H_2O
  $(CH_3CO)_2O + H_2O \rightarrow 2CH_3COOH$
  making ethanoic acid — a carboxylic acid.
- With methanol, CH_3OH
  $(CH_3CO)_2O + CH_3OH \rightarrow CH_3COOCH_3 + CH_3COOH$
  making methyl ethanoate — an ester.
- With ammonia, NH_3
  $(CH_3CO)_2O + 2NH_3 \rightarrow CH_3CONH_2 + CH_3COONH_4$
  making ethanamide — an amide.
- With methylamine, CH_3NH_2:
  $(CH_3CO)_2O + CH_3NH_2 \rightarrow CH_3CONHCH_3 + CH_3COOH$
  making an N-methylethanamide.

Now test yourself

5 Draw the structure of the organic product formed when the compound below reacts with ethanoic anhydride, $(CH_3CO)_2O$.

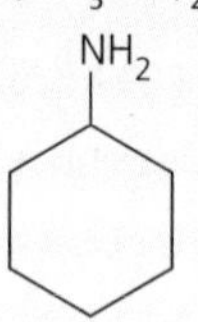

Answers on p. 109

Tested

Exam practice

1 **(a)** Give the molecular formula of propanal. [1]

(b) Draw the structure of the organic product formed when propanal reacts with:

(i) $NaBH_4(aq)$ [1]

(ii) $AgNO_3(aq)$ and $NH_3(aq)$ (Tollens' reagent) [1]

(iii) HCN with a trace of KCN [1]

(c) The product from the reaction in part (b)(iii) is chiral. Draw the enantiomeric forms of the product. [2]

2 Compounds A–C have structural formulae.

A B C

(a) Describe a chemical test that would enable you to distinguish between compounds A and B. [2]

(b) Write an equation, using [H], to show how compound B reacts with $NaBH_4$. [2]

(c) Draw the structure of the organic compound formed when C reacts with Fehling's solution. [1]

3 Give the reagents and conditions required for the chemical transformations:

(a) Y to X [2]

(b) Y to Z [2]

X Y Z

Answers and quick quizzes online

Online

Examiners' summary

You should now have an understanding of:

- ✔ the functional groups in aldehydes and ketones
- ✔ the carbonyl group and how it is polarised
- ✔ how to distinguish between aldehydes and ketones
- ✔ the reactions of the carbonyl group
- ✔ the mechanism of nucleophilic addition
- ✔ carboxylic acids and how they react as acids
- ✔ how carboxylic acids react with alcohols in the presence of an acid catalyst to form an ester
- ✔ how esters can be hydrolysed either in acidic or basic conditions
- ✔ what is meant by 'biodiesel'
- ✔ the acylation reaction
- ✔ the reactions of acyl chlorides and acid anhydrides as typical acylating agents
- ✔ the mechanism by which acyl chlorides react — the nucleophilic addition–elimination mechanism

6 Aromatic compounds

Structure of benzene

Benzene is an unsaturated cyclic hydrocarbon with molecular formula C_6H_6. The molecule has a planar hexagonal structure. It consists of a σ-bonded framework in which all the H–C–H angles are 120°. There is also a delocalised π-electron system above and below the plane of atoms. The π system is formed by the overlap of p orbitals on adjacent carbon atoms, as shown in Figure 6.1.

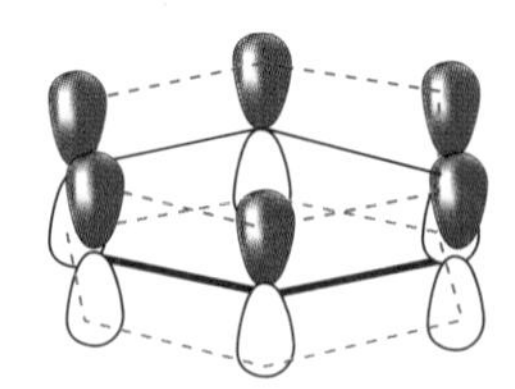

Figure 6.1 Delocalised π electrons above and below the plane of the benzene ring

- The p system above and below the benzene ring consists of six p electrons that are delocalised. This enhances the thermodynamic stability of the ring.
- The carbon–carbon bonds in benzene are intermediate in length between those of a single bond and a double bond (C–C = 0.154 nm; C=C = 0.134 nm; in benzene = 0.139 nm). This is evidence of delocalisation in that intermediate-length bonds are formed.
- The internal bond angles in the benzene ring are 120°. All six carbon atoms have a trigonal planar arrangement — three σ bonds are formed by each and the fourth electron of each carbon is in the delocalised π cloud.

Examiners' tip

When discussing the structure of benzene, refer to the overlapping of p orbitals to form the σ bonds to hydrogen atoms and the delocalised π cloud of electrons.

The benzene ring is best represented as shown in Figure 6.2.

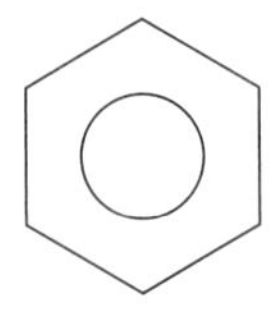

Figure 6.2 Representation of the benzene ring

although it is sometimes convenient to consider the Kekulé form shown in Figure 6.3.

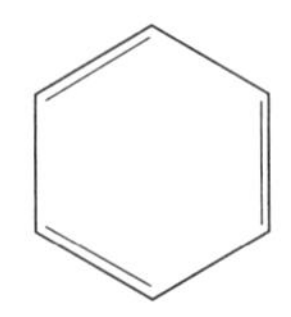

Figure 6.3 The Kekulé form

The Kekulé form is not strictly correct because it implies that separate double and single bonds exist in the molecules. We now know this not to be the case and that delocalisation takes place (see the evidence above).

Notice that the hydrogen atoms are not normally indicated in these structures.

Benzene stability

Revised

Benzene is more thermodynamically stable than expected. The lowering of its energy due to the delocalisation can be quantified by using a simple energy level diagram (Figure 6.4).

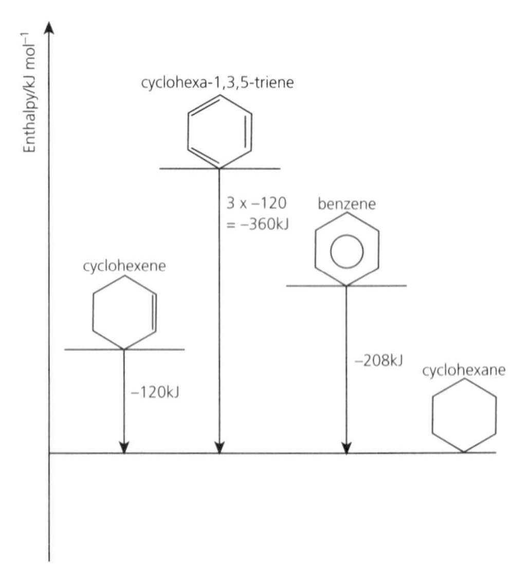

Figure 6.4 Hydrogenation enthalpy evidence for benzene's structure

In all the reactions in Figure 6.4, a process of hydrogenation is represented — a simple addition reaction of hydrogen with three substances:

- 'real' benzene — in which delocalisation is assumed. The hydrogenation enthalpy is $-208\,kJ\,mol^{-1}$

  $C_6H_6(g) + 3H_2(g) \rightarrow C_6H_{12}(g)$
- Cyclohexene — a cyclic alkene that has only one carbon–carbon double bond. Its enthalpy of hydrogenation is $-120\,kJ\,mol^{-1}$:

  $C_6H_{10}(g) + H_2(g) \rightarrow C_6H_{12}(g)$
- Cyclohexa-1,3,5-triene — a 'theoretical' compound that has three carbon–carbon double bonds and no delocalisation is assumed. The hydrogenation enthalpy is estimated to be three times the value for cyclohexene, that is: $-120 \times 3 = -360\,kJ\,mol^{-1}$:

  $C_6H_6(g) + 3H_2(g) \rightarrow C_6H_{12}(g)$

It can be seen from Figure 6.4 that the 'real' benzene is $152\,kJ\,mol^{-1}$ lower in energy than the 'theoretical' benzene molecule. This lowering in energy, due to the delocalised electrons, explains why the reactions of benzene have higher activation energies than expected. They are therefore slower and require higher temperatures and more 'severe' conditions than the reactions of a typical alkene.

Examiners' tip

Try to remember the evidence for delocalisation in benzene — data regarding bond lengths, bond angles and enthalpy of hydrogenation are all useful to know.

Reactions of benzene

Most of the reactions of benzene are called **electrophilic substitution** processes. In these reactions, one or more of the hydrogen atoms in benzene is/are substituted by other group(s).

Nitration of the benzene ring

Revised

Nitrating a benzene ring involves placing a nitro group ($-NO_2$) into benzene instead of a hydrogen — the process is called **nitration**. Nitrobenzene has the structure shown in Figure 6.5.

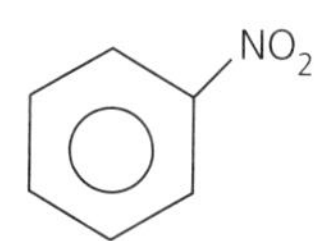

Figure 6.5 Nitrobenzene

The reagents for carrying out **nitration** are concentrated nitric(v) acid, $HNO_3(l)$, a catalytic amount of concentrated sulfuric(vi) acid, $H_2SO_4(l)$, at a maximum temperature of 50°C.

An equation for the reaction is shown in Figure 6.6.

$$C_6H_6 + HNO_3 \xrightarrow[50°C]{H_2SO_4} C_6H_5NO_2 + H_2O$$

Figure 6.6 Nitration of benzene

Electrophilic substitution mechanism

An electrophile called the nitronium ion, NO_2^+, is produced:

$$HNO_3 + H_2SO_4 \rightleftharpoons HSO_4^- + H_2NO_3^+$$

$$H_2NO_3^+ \rightarrow NO_2^+ + H_2O$$

Examiners' tip

It is important to know the electrophilic substitution mechanism — there are several variations on it that are very similar. Remember and know one of them, and the others can often be worked out.

The electrophile then reacts with the benzene ring (Figure 6.7).

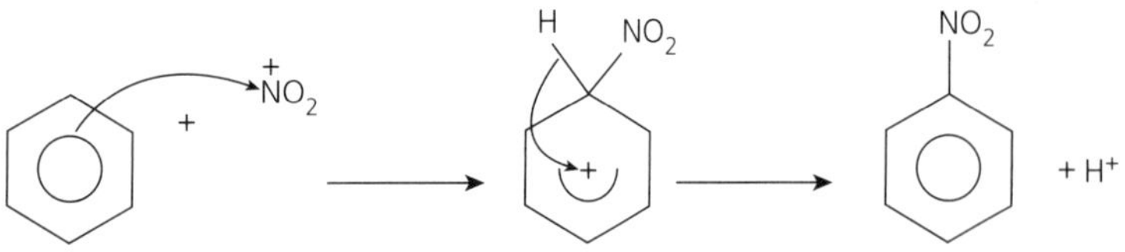

Figure 6.7 Electrophilic substitution — nitration

- Two electrons are donated from the π cloud of delocalised electrons in the benzene ring.
- A new carbon–nitrogen bond forms in an intermediate species. The benzene ring in this has only partial delocalisation — this raises its energy.
- A carbon–hydrogen bond breaks and the two electrons in the bond move back into the π system on the benzene ring to restore the delocalisation.
- The resulting proton, H^+, recombines with the hydrogensulfate(VI) ion, HSO_4^-, to regenerate the H_2SO_4 catalyst.

Uses of nitration

Many of the nitro products that can be formed by this reaction are used for making explosives — for example, 2-methyl-1,3,5-trinitrobenzene (Figure 6.8), otherwise known as 2,4,6-trinitrotoluene (TNT).

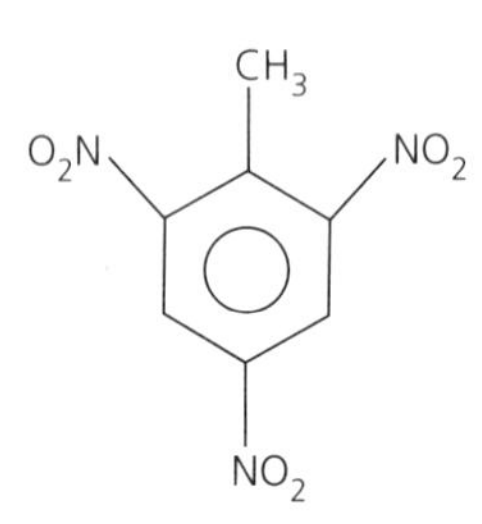

Figure 6.8 TNT

Reduction of these aromatic nitro compounds is achieved by refluxing with tin in concentrated hydrochloric acid — this yields the corresponding aromatic amine (see Chapter 7).

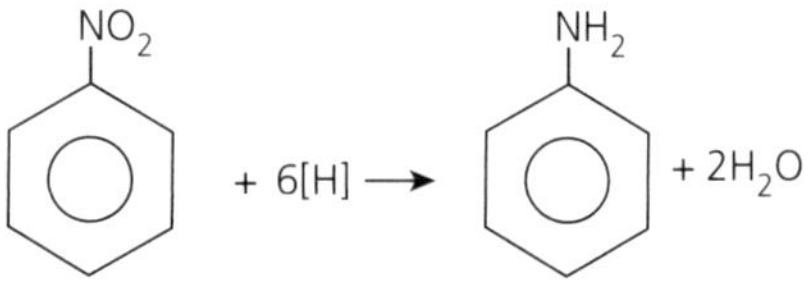

The phenylamine formed using this reaction (Figure 6.9) can be used for making **dyestuffs**.

Figure 6.9 Making phenylamine

Friedel–Crafts acylation

Revised

As stated previously, **acylation** involves incorporating an RCO group into a molecule. Figure 6.10 shows the reaction between benzene and ethanoyl chloride in the presence of aluminium chloride, $AlCl_3$.

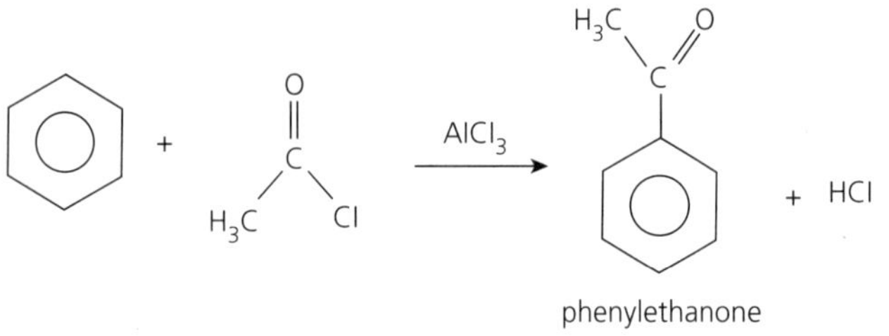

Figure 6.10 Making phenylethanone

The mechanism for the reaction is **electrophilic substitution**.

The required electrophile (the acylium ion), CH_3C^+O, is formed using an aluminium chloride catalyst, $AlCl_3$ (Figure 6.11).

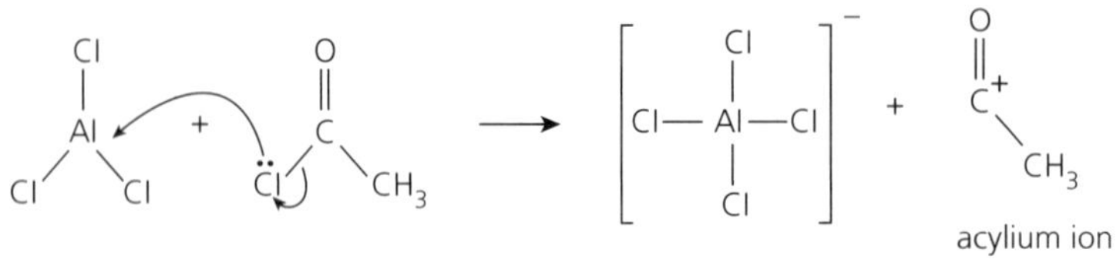

Figure 6.11 Making the electrophile

Two π electrons from the benzene ring are donated to the positive carbon of the acylium ion and a new C–C bond is formed (Figure 6.12). Loss of a proton from the intermediate then regenerates the delocalised π cloud in the benzene ring.

Figure 6.12 The substitution step

The released proton can reacts with the $AlCl_4^-$ ion to regenerate the catalyst:

$$AlCl_4^- + H^+ \rightarrow AlCl_3 + HCl$$

Examiners' tip

This mechanism is virtually the same as the nitration mechanism covered earlier, except that the electrophile is CH_3C^+O instead of N^+O_2.

Now test yourself Tested

1 Examine the reactions below.

Reaction 1:

Reaction 2:

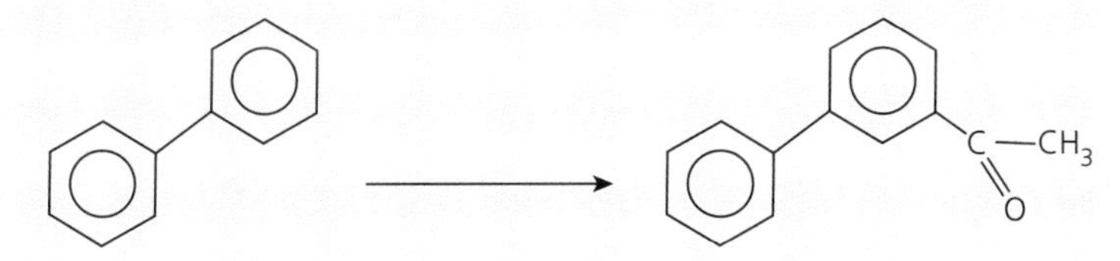

(a) Give the formulae of the electrophile featured in each reaction.

(b) Give the reagents and conditions required for each reaction.

Answers on p. 109

Revision activity

As the number of reactions you need to know starts to build up, produce spider diagrams or reaction schemes (in colour) and keep looking at them to try to remember the reactions.

Exam practice

1 The questions below relate to this reaction scheme:

A B C D

(a) Give the reagents required for each stage in the reaction sequence. [3]

(b) Explain why a racemic mixture forms when C is synthesised. [2]

(c) Suggest the name of a reagent that could be used to convert C back into B. [1]

(d) Name and outline the mechanism for the formation of B from A. [3]

2 The following scheme enables compound M to be synthesised from benzene.

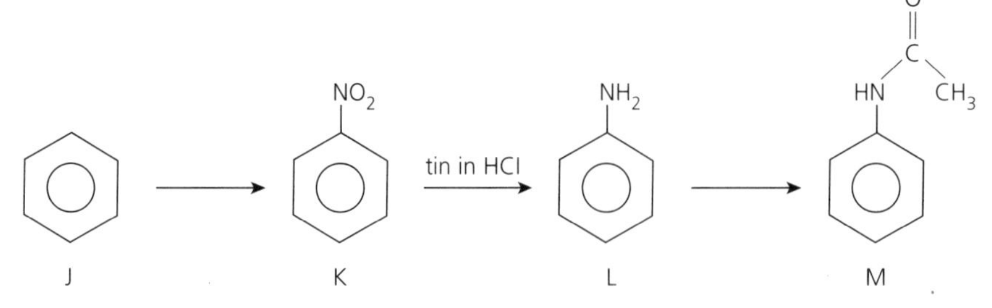

(a) What is the molecular formula for compound M? [1]

(b) What type of reaction takes place when K is converted to L? [1]

(c) Suggest reagents and conditions required for the conversion from:

(i) J to K [1]

(ii) L to M [1]

(d) Write an equation for the conversion of J into K. [1]

(e) Give the names of the mechanisms involved in the conversion from:

(i) J to K [1]

(ii) L to M [1]

(f) Suggest a use for compounds that can be formed using the type of reaction when J is converted into K. [1]

Answers and quick quizzes online

Online

Examiners' summary

You should now have an understanding of:

- ✔ the structure of benzene in terms of the π cloud of delocalised electrons and how this is formed by atomic orbital overlap
- ✔ the evidence to support the idea of electron delocalisation in benzene rings
- ✔ nitration as an important example of a reaction involving benzene
- ✔ the electrophilic substitution mechanism
- ✔ Friedel–Crafts acylation and its mechanism
- ✔ the use of nitration and Freidel–Craft acylation in organic syntheses

7 Amines and amino acids

Amines

Amines can be thought of as alkylated ammonia species in which ammonia's hydrogen atoms are substituted by one or more alkyl groups.

Amine molecules, like ammonia, are pyramidal in their spatial arrangement of covalent bonds around the nitrogen atom (Figure 7.1) with the internal angle being approximately 107°.

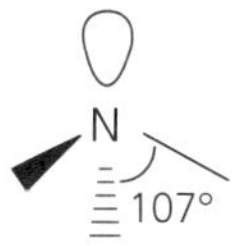

N 107°

Figure 7.1 Spatial arrangement around a saturated nitrogen atom

Amines are classed according to the number of alkyl groups (or C–N) bonds surrounding the central nitrogen atom.

- Primary amines have 1 attached alkyl group:

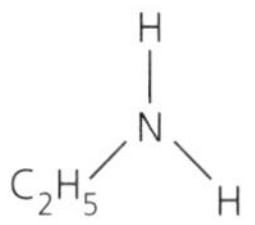

H, N, C_2H_5, H

ethylamine

- Secondary amines have 2 attached alkyl groups:

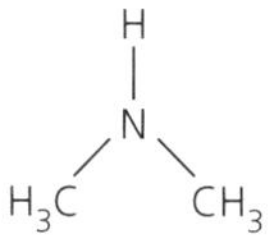

H, N, H_3C, CH_3

dimethylamine

- Tertiary amines have 3 attached alkyl groups:

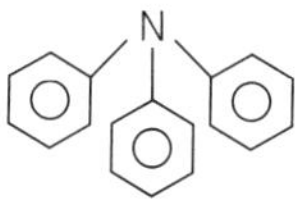

N

triphenylamine

- Quaternary ammonium salts have 4 attached alkyl groups:

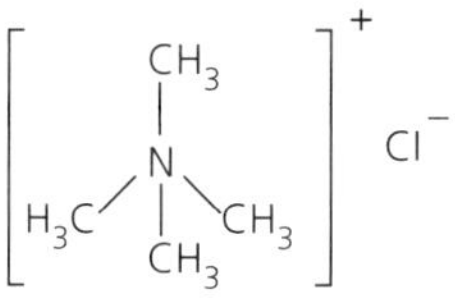

$[N(CH_3)_4]^+ Cl^-$

tetramethylammonium chloride

Amines as bases

Revised

Phenylamine — an aromatic amine

Phenylamine (Figure 7.2) consists of an amino group ($-NH_2$) attached directly to a benzene ring.

Nitrogen's lone pair of electrons can overlap with the π-system in the benzene ring and, as a result, the lone pair of the nitrogen atom is delocalised into the π-system.

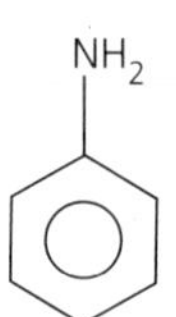

Figure 7.2 Phenylamine

Phenylamine is a weaker Brönsted–Lowry base than ammonia because nitrogen atom's lone pair is less available for protonation:

$$C_6H_5NH_2 + H^+ \rightarrow C_6H_5NH_3^+$$

Aliphatic amines, in which the alkyl groups are attached directly to the nitrogen atom, have a slightly enhanced electron density on the nitrogen atom due to the electron-donating properties of the alkyl groups. This makes the nitrogen lone pair easier to protonate than in ammonia (no alkyl groups) and much easier than in phenylamine.

Examiners' tip

When explaining the relative basic strength of amines, use the correct terms in your explanation — Brønsted–Lowry base, lone pair, delocalised, protonation, π-system, electron density etc.

Aliphatic amines generally increase in base strength as the number of alkyl groups attached to the nitrogen atom increases. Each alkyl group donates electrons towards the nitrogen atom, so the more alkyl groups there are, the greater will be the electron density on the nitrogen, and the easier protonation will take place.

The order, with strongest base first, is:

aliphatic amines > ammonia > phenylamine

Amines as nucleophiles

Revised

Amines, like ammonia, are nucleophiles and so react with haloalkanes (by nucleophilic substitution) and with acyl chlorides (by nucleophilic addition–elimination) — see page 37.

Amines reacting with haloalkanes

Haloalkanes have a polar carbon–halogen bond ($^{\delta+}C– X^{\delta-}$) and this is attacked by the nitrogen atom in amine molecules displacing the halogen as a halide ion.

An excess of ammonia (in ethanol solvent) can react with bromoethane, C_2H_5Br:

$$C_2H_5Br + NH_3 \rightarrow C_2H_5NH_2 + HBr$$

but preferably:

$$C_2H_5Br + 2NH_3 \rightarrow C_2H_5NH_2 + NH_4Br$$

This is because the acidic HBr and any remaining ammonia in the reaction mixture react immediately to form ammonium bromide.

Iodomethane reacts with ethylamine (a primary amine):

$$CH_3I + C_2H_5NH_2 \rightarrow C_2H_5NHCH_3 + HI$$

forming ethylmethylamine — a secondary amine. Depending on conditions, the protonated amine iodide salt $[C_2H_5NH_2CH_3]^+ I^-$ may form.

If excess iodomethane is present this process can continue producing a mixture of products (Figure 7.3). Secondary amines, tertiary amines and quaternary ammonium salts may all be present depending on the amount of iodomethane present.

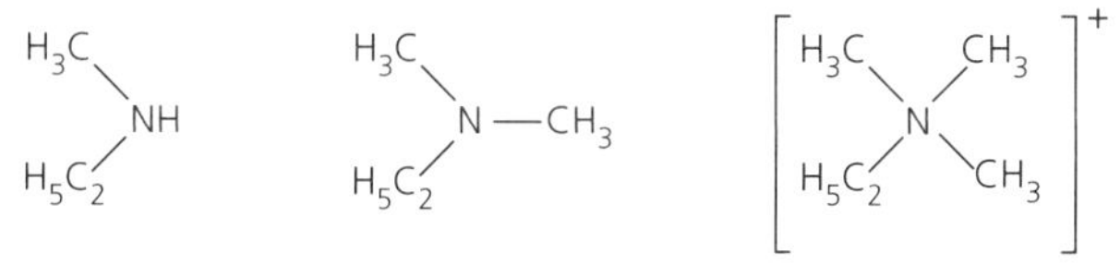

secondary amine | tertiary amine | quaternary ammonium ion

Figure 7.3 Successive reaction products

Quaternary ammonium salts are used as **cationic surfactants** in detergents. Examples of two of these are shown in Figure 7.4. The polar/charged (hydrophilic) nature of one end of the ion and the hydrophobic nature of the hydrocarbon chain can be exploited in their action as detergents, particularly in removing unwanted grease.

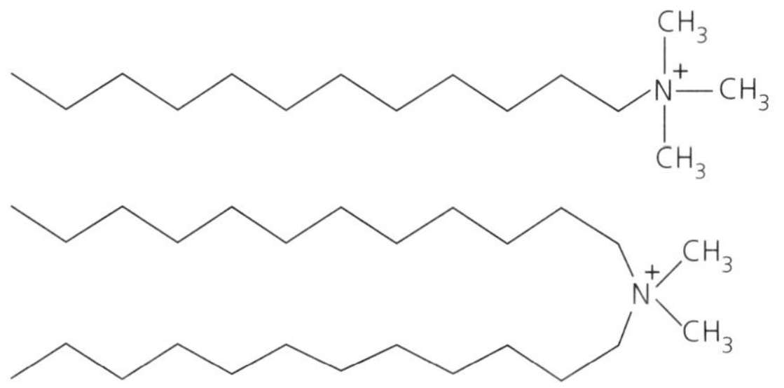

Figure 7.4 Detergents

Nucleophilic substitution mechanism

In the reaction between a bromoalkane and ammonia (Figure 7.5):

- the ammonia molecule acts as a **nucleophile** by donating its lone pair of electrons to the slightly positively charged carbon atom in the $^{\delta+}C–Br^{\delta-}$ bond.
- A new carbon–nitrogen bond forms.
- An ammonia molecule (acting as a base) then removes a proton from the $–NH_3^+$ group forming an amine.

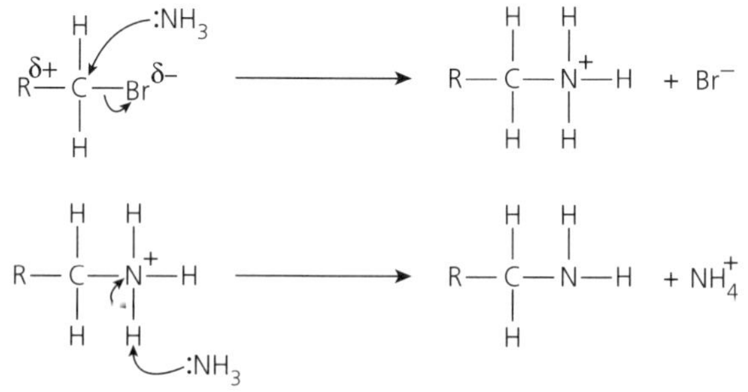

Figure 7.5 Nucleophilic substitution

Examiners' tip

This mechanism is similar to the AS mechanism in which hydroxide ions, OH^-, act as nucleophiles when attacking a halogenoalkane. The difference is the need for a second ammonia molecule to remove a proton in the amine reaction.

Examiners' tip

The rate of this reaction depends on the strength of the C–X bond — iodoalkanes react faster than bromoalkanes because a C–I bond is weaker than a C–Br bond.

Synthesis of amines

Revised

Aliphatic amines

Nitriles are organic molecules that contain a CN group. They can be reduced to primary amines using hydrogenation, with the help of a nickel catalyst:

$$CH_3CN + 2H_2 \rightarrow CH_3CH_2NH_2$$

ethylamine

Nitriles can also be reduced using **sodium tetrahydridoborate(III), $NaBH_4$**, in a non-aqueous solvent, like ethanol, to form primary amines:

$$C_2H_5CN + 4[H] \rightarrow C_2H_5CH_2NH_2$$

propylamine

This reaction is also reduction and a balanced equation can be written using [H] to represent the role of the $NaBH_4$.

Aromatic amines

Aromatic amines such as phenylamine (Figure 7.6) can be prepared by reducing the appropriate aromatic nitro compound by refluxing with tin in concentrated hydrochloric acid.

Figure 7.6 Preparation of phenylamine

Now test yourself

1 What are the products of the following reactions?

(a)

(b)

Answers on p. 109

Tested

Amino acids

The general formula of an α-amino acid is $H_2NC(R)(H)COOH$ and Figure 7.7 shows its structure.

- The α-carbon atom is the one that is linked to both the COOH and NH_2 group — all naturally occurring amino acids have this feature.
- The R group can vary but when it is a hydrogen atom, as in glycine, the molecule is not chiral because two of the four groups attached to the central (*) carbon atom are identical. However, all other amino acids are chiral and can exist in either the (+) or (–) enantiomeric forms. It is an interesting feature of their stereochemistry that all naturally occurring optically active amino acids are of the 'left-hand' variety or (–) form. This means that they all rotate the plane of polarised light to the left.
- The enantiomeric forms of a general amino acid are shown in Figure 7.8 — note their mirror image relationship to each other.

Figure 7.7 General formula for an α-amino acid

Figure 7.8 Chirality in α-amino acids

Alanine and serine (Figure 7.9) are further examples of naturally occurring amino acids — both are chiral (unlike glycine).

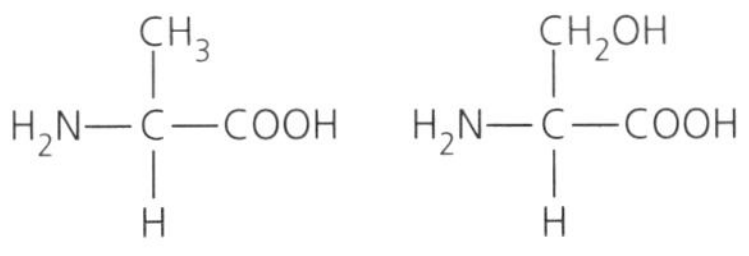

Figure 7.9 Two simple α-amino acids

Acid–base reactions of amino acids

Amino acids have both a basic amine group (NH_2) and an acidic carboxylic acid group (COOH). When an acid or a base is added to an amino acid, the base or acid group reacts, as illustrated by alanine in Figure 7.10.

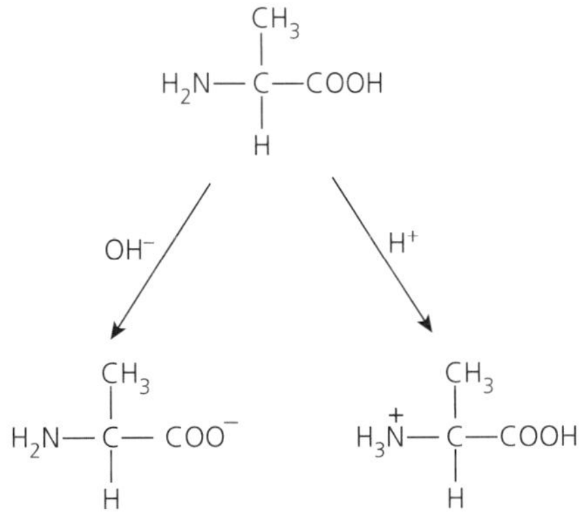

Figure 7.10 Alanine acting as a base and an acid

Examiners' tip

Learn these reactions — they are easy to understand. The NH_2 group is protonated in acids, whereas the COOH group loses a proton in a basic environment.

An **isoelectronic point** is the pH at which a **zwitterion** of an amino acid exists — the overall charge being zero At an isoelectronic point both the NH_3^+ and COO^- groups are present. Solid amino acids have a higher than expected melting points because of the presence of zwitterions.

Proteins

Amino acids are able to combine with each other, in a condensation process, to form **proteins**. Amino acids are joined to each other by **peptide links**.

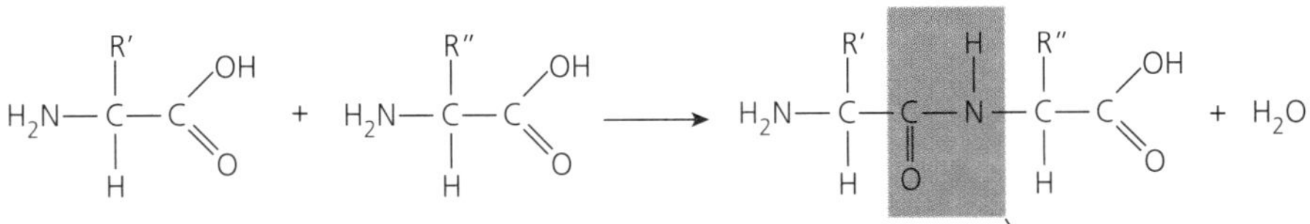

Figure 7.11 Protein formation

The short protein strand shown in Figure 7.11 consists of two amino acid 'residues' and is called a **dipeptide**.

It is possible to **hydrolyse** a protein, either in acidic or basic conditions, to form the constituent amino acids. Figure 7.12 shows the hydrolysis of a tripeptide — a short protein strand consisting of three amino acid 'residues' — to form three amino acids.

tripeptide

peptide links

$$H_2N{-}CHR'{-}CO{-}NH{-}CHR''{-}CO{-}NH{-}CHR'''{-}COOH$$

hydrolysis

amine group

carboxyl group

$$H_2N{-}CHR'{-}COOH + H_2N{-}CHR''{-}COOH + H_2N{-}CHR'''{-}COOH$$

amino acids (side chains are R', R'' and R''')

Figure 7.12 Hydrolysis of a protein

The amino acids formed can be separated by **chromatography**. Analysis of a chromatogram can reveal the structure of the original, although it is a highly complex procedure.

Hydrogen bonding in proteins

Revised

Protein chains can interact with other protein chains using hydrogen bonds. Figure 7.13 shows the interaction between a carbonyl group, $^{\delta+}C{=}O^{\delta-}$, in one peptide link with the $^{\delta-}N{-}H^{\delta+}$ group of another peptide link in another protein strand.

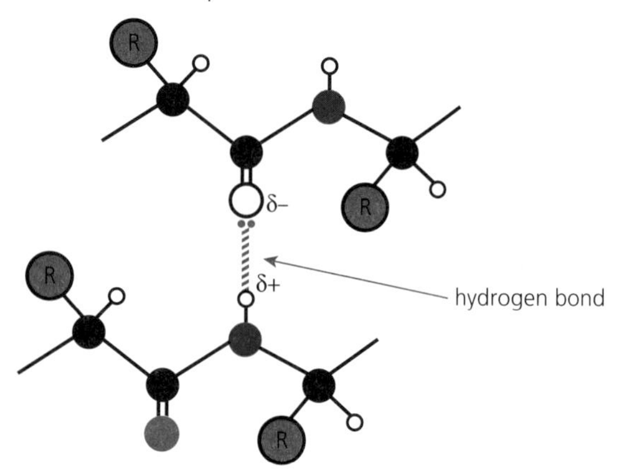

Figure 7.13 Hydrogen bonding in proteins

β-pleated sheets and the α-helix are two of several 3-dimensional forms of proteins that owe their structures to the existence of hydrogen bonds.

Now test yourself

2 The two amino acids shown below are commonly called alanine and aspartic acid.

$$H_3C{-}CH(NH_2){-}COOH$$

alanine

$$HOOC{-}CH_2{-}CH(NH_2){-}COOH$$

aspartic acid

Draw the structures of:

(a) the zwitterion of alanine

(b) the alanine species that would exist at pH 3

(c) three dipeptides that could be formed by alanine reacting with aspartic acid.

Answers on p. 109

Tested

Exam practice

1 Consider the three compounds drawn below, all of which are Brønsted–Lowry bases.

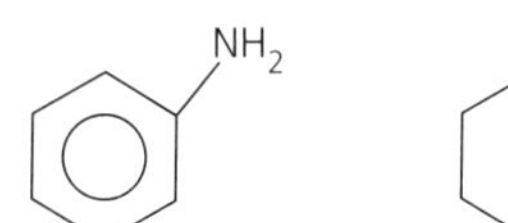

NH_2 NH_3

phenylamine cyclohexylamine ammonia

(a) State what is meant by the term 'Brønsted–Lowry base'. [1]

(b) Write an equation to show cyclohexylamine acting as a base. [1]

(c) Write the compounds in order of decreasing basic strength — strongest base first. [1]

(d) Explain your answer to part (c). [3]

(e) Draw the structure of the organic compound formed when phenylamine reacts with:

(i) bromomethane, CH_3Br [1]

(ii) ethanoic anhydride, $(CH_3CO)_2O$ [1]

2 The structure below shows a dipeptide formed by alanine and glycine. The shorthand form of this structural sequence is ala–gly; that is an alanine unit bonded to a glycine unit. It is shown in its zwitterionic form.

CH_3 H O H C N C C H_3N^+ H C C O^- O H

ala-gly

(a) Draw the structure of the dipeptide called 'gly–ala'. [1]

(b) Are the structures of gly–ala and ala–gly the same? Explain your answer. [2]

(c) Draw the structures of the hydrolysis products formed when ala–gly is warmed with:

(i) NaOH(aq) [2]

(ii) HCl(aq) [2]

Answers and quick quizzes online

Online

Examiners' summary

You should now have an understanding of:

- ✔ how amines can act as Brønsted–Lowry bases
- ✔ the structural features that determine base strength
- ✔ how amines react with halogenoalkanes
- ✔ how aliphatic amines can be prepared by reducing a nitrile
- ✔ how aromatic amines can be prepared by reduction of a nitro compound
- ✔ amino acids and their general structure
- ✔ zwitterions and how they are formed
- ✔ proteins and how they are formed from amino acids
- ✔ protein hydrolysis and how different products form depending on the pH at which the hydrolysis takes place

8 Polymers

Polymers are formed when many small molecules, or units called **monomers**, bond together to form long chains. There are two types of polymer that can form depending on the original molecules — addition polymers and condensation polymers.

Addition polymers

Molecules containing carbon–carbon double bonds can be polymerised under high pressure and high temperature conditions using a range of catalysts to form **addition polymers**. The π bonding electrons from the double bond are used to join carbon atoms together with the result that an extensive carbon chain is formed (Figure 8.1). This chain constitutes the backbone of the polymer.

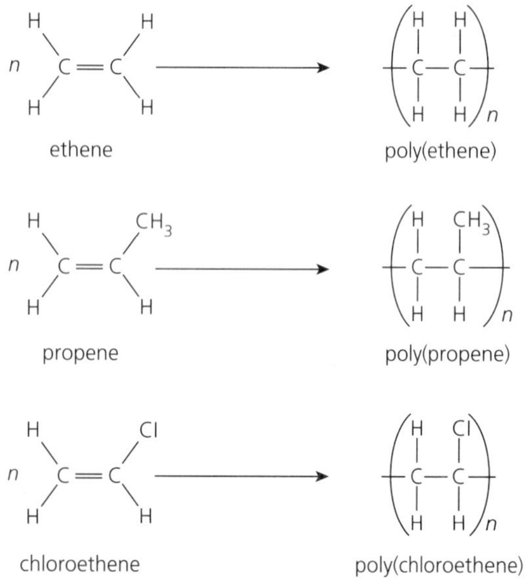

$$n\,H_2C{=}CH_2 \longrightarrow {-}\!\left(CH_2{-}CH_2\right)_n\!{-}$$

ethene → poly(ethene)

$$n\,H_2C{=}CH(CH_3) \longrightarrow {-}\!\left(CH_2{-}CH(CH_3)\right)_n\!{-}$$

propene → poly(propene)

$$n\,H_2C{=}CHCl \longrightarrow {-}\!\left(CH_2{-}CHCl\right)_n\!{-}$$

chloroethene → poly(chloroethene)

Figure 8.1 Addition polymerisation

If the structure of a polymer is given in a question, the monomer can be deduced by spotting the repeating unit (Figure 8.2), and then drawing out the monomer with its carbon–carbon double bond shown.

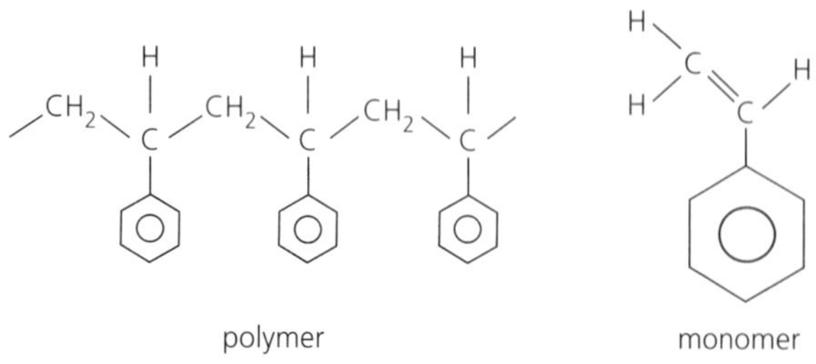

polymer: $-CH_2-CH(C_6H_5)-CH_2-CH(C_6H_5)-CH_2-CH(C_6H_5)-$

monomer: $H_2C{=}CH(C_6H_5)$

Figure 8.2 What is the monomer?

Now test yourself Tested

1 Draw the structures of the monomers that would polymerise to form the addition polymers:

(a) $-CH(CONH_2)-CH_2-CH(CONH_2)-CH_2-CH(CONH_2)-CH_2-CH(CONH_2)-CH_2-CH(CONH_2)-CH_2-$

(each CH carries a side group $C{=}O$ bonded to NH_2)

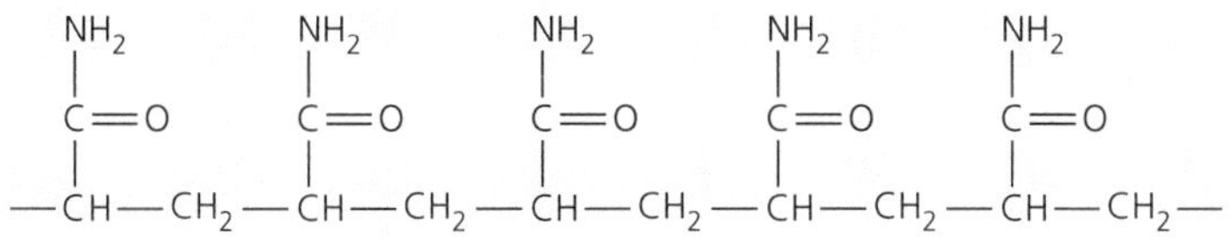

(b) $-CH(OCOCH_3)-CH_2-CH(OCOCH_3)-CH_2-CH(OCOCH_3)-CH_2-CH(OCOCH_3)-CH_2-CH(OCOCH_3)-CH_2-$

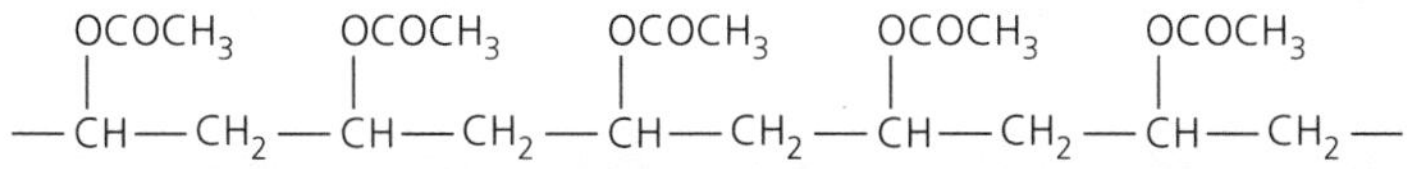

2 Draw the repeating unit of the polymer expected when this monomer polymerises.

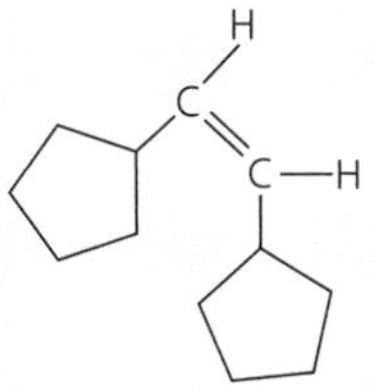

H—C(cyclopentyl)=C—H(cyclopentyl)

Answers on p. 109

Condensation polymers

Some molecules that have more than one functional group are able to link together and form long chains. A small molecule like water or hydrogen chloride is also formed in this type of polymerisation and the polymer, as a result, is called a **condensation polymer**.

Nylon (a polyamide) and **terylene** (a polyester) are examples of condensation polymers. When forming nylon (Figure 8.3) it is possible to use a dicarboxylic acid and a diamine.

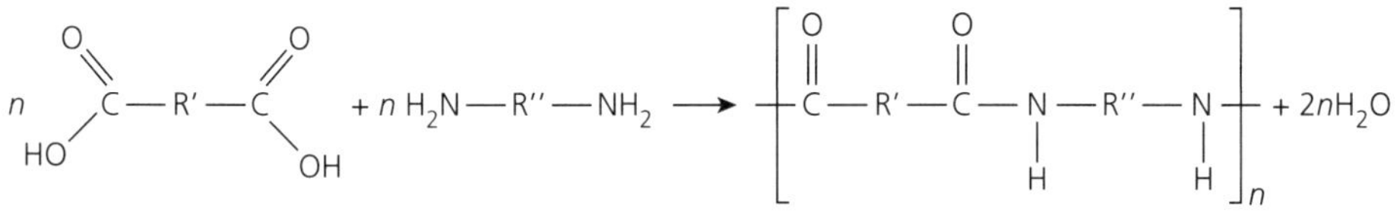

$$n\,HOOC-R'-COOH + n\,H_2N-R''-NH_2 \longrightarrow \left[-\overset{O}{\overset{\|}{C}}-R'-\overset{O}{\overset{\|}{C}}-\underset{H}{\underset{|}{N}}-R''-\underset{H}{\underset{|}{N}}-\right]_n + 2nH_2O$$

Figure 8.3 Condensation polymerisation making nylon

Terylene is a polyester and is made (Figure 8.4) by the reaction of benzene-1,4-dicarboxylic acid with ethane-1,2-diol.

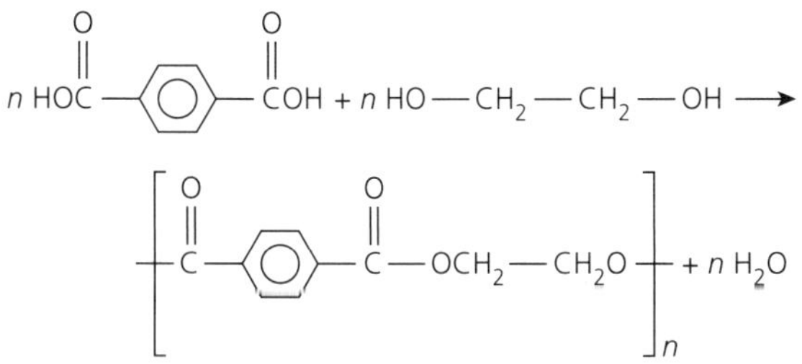

$$n\,HO\overset{O}{\overset{\|}{C}}-C_6H_4-\overset{O}{\overset{\|}{C}}OH + n\,HO-CH_2-CH_2-OH \longrightarrow$$

$$\left[-\overset{O}{\overset{\|}{C}}-C_6H_4-\overset{O}{\overset{\|}{C}}-OCH_2-CH_2O-\right]_n + n\,H_2O$$

Figure 8.4 Condensation polymerisation making terylene

Kevlar is a condensation polymer (a polyamide) that has been put to many interesting uses — for example in making bulletproof vests, in which strength and resistance to rapid extreme forces are properties that make it particularly useful. It is made by reacting two monomers — a diacycl chloride and a diamine.

The interactions between two strands of Kevlar are shown in Figure 8.5. Notice how hydrogen bonds bind different strands of polymer together.

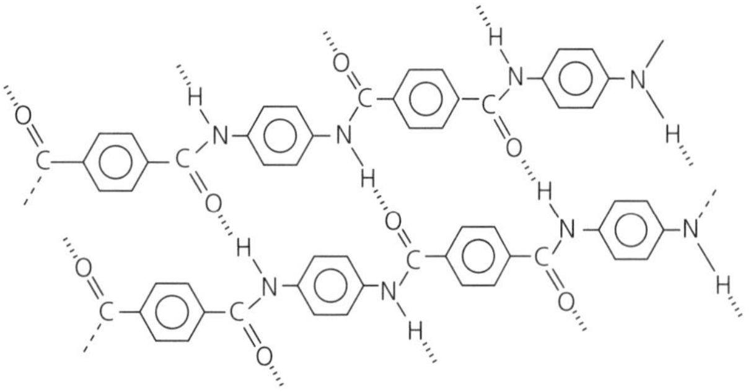

Figure 8.5 Kevlar

Now test yourself

Tested

3 This is the structure of a condensation polymer.

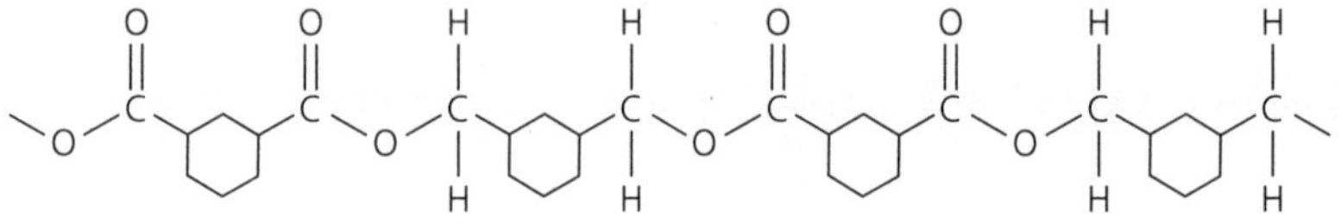

Draw the structures of the two monomers that would form this polymer.

4 The structure for the compound called phenylalanine is drawn below.

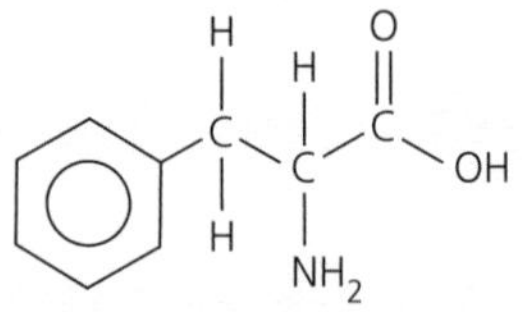

(a) What type of compound is phenylalanine?

(b) Draw the structure of the repeating unit that phenylalanine forms when polymerised.

(c) It is possible for phenylalanine to form a cyclic dimer in which two molecules combine, also producing two molecules of water. Draw the structure of this dimer.

Answers on p. 110

Disposing of polymers

When objects made of polymers come to the end of their useful lives, they are often disposed of in landfill waste sites. There are serious issues associated with this.

- Many addition polymers, like polyalkenes, are **chemically inert** — they contain strong carbon–carbon bonds and are non-polar. This means that possible nucleophiles, like water, will not decompose (**biodegrade**) these polymers even in acidic or basic conditions. As a result, many addition polymers have a long residence time during which they will only biodegrade over many centuries.
- Polar polymers, like polyesters and various nylons, are biodegradable and they are hydrolysed to form their constituent monomers, many of which are water-soluble.

- Polymers can be burned instead of disposing of them in the ground. The **incineration** process involves high-temperature combustion and will produce heat energy that can then be transformed into electrical energy (using steam turbines). Various pollutants will also form in this process, many of which will be potentially harmful to the environment. For example, toxic and highly acidic hydrogen chloride gas will form if chlorinated polymers like polyvinylchloride are incinerated. Incineration also produces hazardous compounds called dioxins, for example polychlorinated biphenyls (PCBs) — but modern incinerators can now reach very high temperatures which destroy these dioxins.
- Polymers that melt on warming (thermoplastic polymers) can be remoulded and new products can be produced at a considerably lower energy cost than forming a polymer from crude oil via cracking etc. Sorting polymers prior to melting can be problematic because many different polymers are used.

Exam practice

1 Propene gas, under certain conditions, will form a waxy, white solid called poly(propene).

(a) State the type of reaction that has taken place. [1]

(b) Give the conditions required for this reaction to take place. [1]

(c) Draw the displayed formula for propene. [1]

(d) Draw a section of a poly(propene) molecule in which **three** propene molecules have joined. [2]

2 Kevlar is a polymer with the structure shown in Figure 8.5.

(a) What type of polymer is Kevlar, addition or condensation? Explain your answer. [2]

(b) Draw the two monomers that can react together to form Kevlar. [2]

3 Identify the compounds mentioned below (V–Z) by drawing their displayed formulae (showing all bonds) and naming them.

(a) Compound V has the molecular formula C_3H_6. It does not react with bromine water and all attempts to polymerise it fail. [1]

(b) The polymer of compound W has this structure: [1]

(c) Compounds X and Y polymerise to form this polymer unit: [2]

(d) Compound Z is formed when the compound shown below is heated with NaOH(aq) and then has HCl(aq) added. [1]

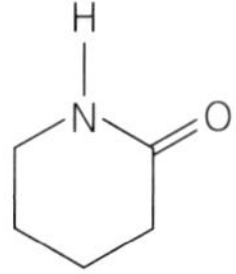

Answers and quick quizzes online

Online

Examiners' summary

You should now have an understanding of:

- ✔ what addition polymers are
- ✔ how to draw the structure of an addition polymer given the monomer
- ✔ how to draw the structure of a monomer given the polymer
- ✔ what condensation polymers are
- ✔ how dicarboxylic acids react with diols to form polyesters like terylene
- ✔ how diamines react with dicarboxylic acids to form polyamides like nylon and Kevlar
- ✔ how polyesters and nylons can be hydrolysed to form their constituent monomers
- ✔ various disposal methods for polymers and how these depend on the chemical properties of the polymers

9 Synthesis, analysis and structure determination

There are many modern analytical techniques that can be used to determine the structures of molecules. These techniques include infrared spectroscopy, nuclear magnetic resonance spectroscopy, mass spectrometry and chromatography.

Infrared spectroscopy

Atoms in molecules are **vibrating** about a fixed position. The frequency of the vibration varies according to the nature of the atoms bonded together:

- the **greater the mass** of the atoms, the lower the frequency of the vibrations
- the **stronger the bond**, the higher the frequency of the vibrations.

The frequency of such vibrations can be quoted either in Hz or as the reciprocal of the wavelength in centimetres (called the **wavenumber** or the number of waves in 1 cm). A particular bond will vibrate or resonate at a particular frequency. The vibration wavenumbers of some covalent bonds is given in Table 9.1.

Table 9.1 Bond vibration wavenumbers

Bond	Wavenumber/cm^{-1}
N–H (amines)	3300–3500
O–H (alcohols)	3230–3550
C–H	2850–3300
O–H (acids)	2500–3000
C≡N	2220–2260
C=O	1680–1750
C=C	1620–1680
C–O	1000–1300
C–C	750–1100

Figure 9.1 shows a typical infrared spectrum.

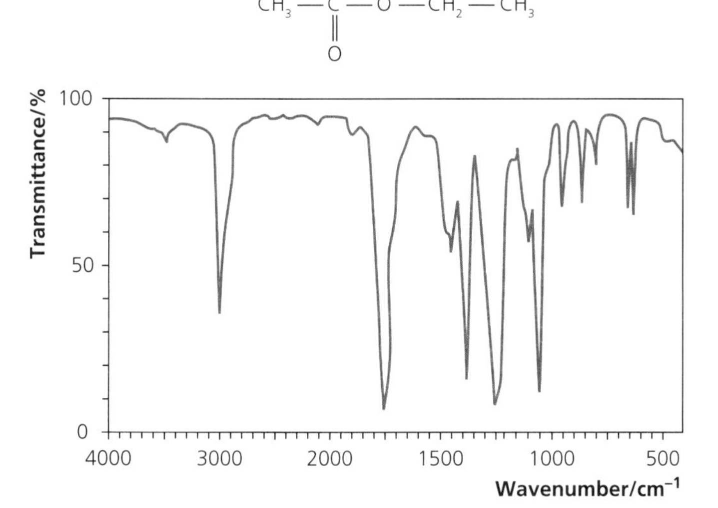

Figure 9.1 IR spectrum of ethyl ethanoate

Note the absorptions taking place at 1740 cm^{-1} (the C=O stretch in the ester) and also at 1240 cm^{-1} (the C–O stretch).

If we didn't know the identity of the substance, we could deduce that a C=O bond and a C–O bond were present, but this is all. However, the fingerprint region (the collection of absorptions below 1500 cm^{-1}) can often be used to identify a compound. This region of the spectrum can be compared to a database of spectra because this **fingerprint region** is unique for a particular molecule.

Mass spectrometry

A sample is injected into a mass spectrometer and a spectrum is produced that measures the **mass-to-charge ratio** (m/z) of a molecule (and its fragmented ions) and plots this against the corresponding **abundance** (the number of ions of a particular mass-to-charge ratio).

A molecule may undergo the following changes in the machine.

- Simple ionisation, in which a high-energy electron is removed from the molecule, M

  $M(g) \rightarrow M^{+\bullet}(g) + e^-$

  $M^{+\bullet}$ is called the **molecular ion**, and an accurate measure of its relative molecular mass can be used to identify the molecule.

  The most useful peak in a mass spectrum is that of the **molecular** ion — seen as the most significant peak on the right of the spectrum.
- The molecular ion may undergo **fragmentation**:

  $M^{+\bullet}(g) \rightarrow X^+(g) + Y^\bullet(g)$

  The charged species formed, X^+, will be detected and its m/z ratio measured.

  Some ions are particularly stable and their abundances tend to be relatively high — these include carbocations (R^+) and acylium ions (RCO^+).

Figure 9.2 shows a typical mass spectrum for, $H_3CCOCH_2OCH_3$

$$H_3C-\overset{\overset{\displaystyle O}{\|}}{C}-CH_2-O-CH_3$$

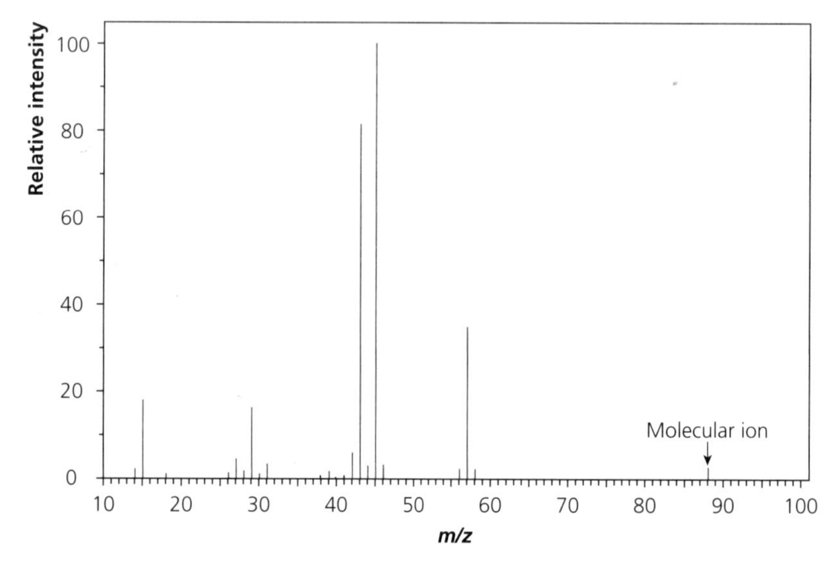

Figure 9.2 Mass spectrum of methoxypropanone

Notice the following about this spectrum:

- The relative molecular mass of the compound is 88 because the peak furthest to the right is at $m/z = 88$
- The peak at $m/z = 43$ is probably due to the fragment ion $[CH_3CO]^+$
- The peak at $m/z = 45$ is probably due to the fragment ion $[CH_3OCH_2]^+$

Now test yourself

Tested

1 The infrared spectrum and mass spectrum for a compound, X, are given below.

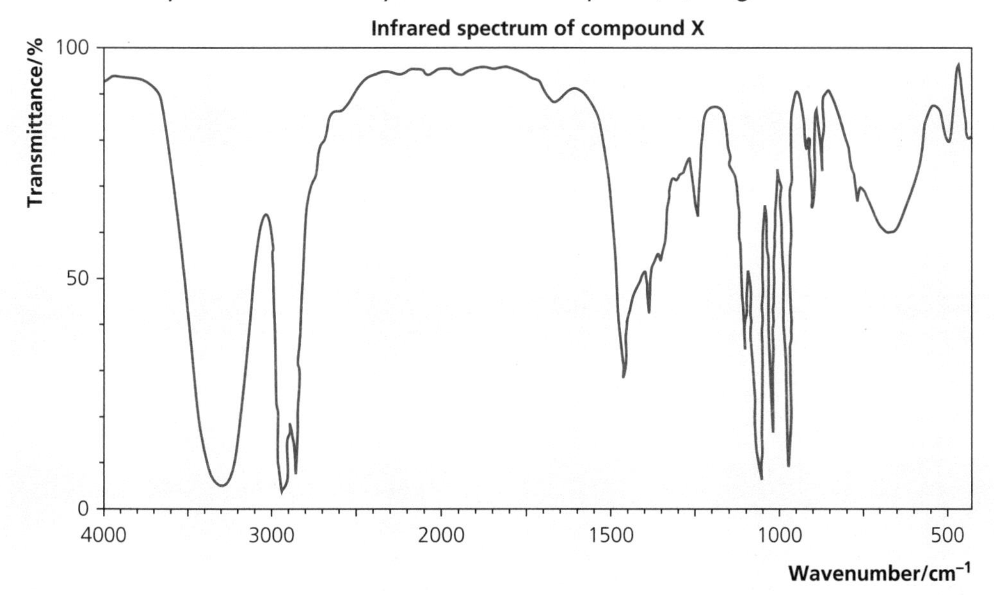

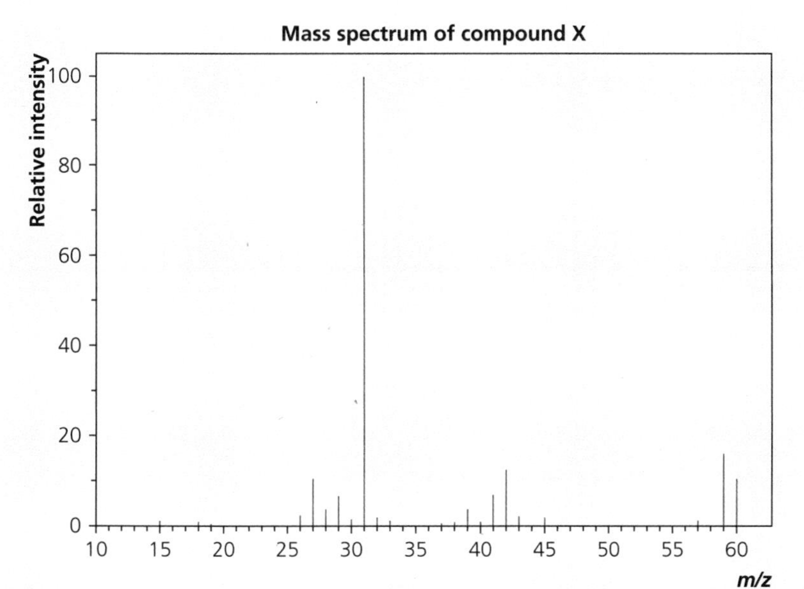

(a) Use the infrared spectrum to deduce which important bonds are present in X.

(b) Use the mass spectrum to determine the relative molecular mass of X.

Given that X contains 60.0% carbon, 13.3% hydrogen and the rest is oxygen, calculate:

(c) The empirical formula of X.

(d) The molecular formula of X.

(e) Suggest a possible structure for X.

Answers on p. 110

Nuclear magnetic resonance spectroscopy

The magnetic properties of some atomic nuclei can be used to determine their environment in molecules that contain them. Absorptions can be detected across the radio wave range of electromagnetic radiation. These can be used to generate a nuclear magnetic resonance (n.m.r.) spectrum.

Nuclei with an odd number of nucleons (protons + neutrons) have a quality known as **nuclear spin**. Examples of nuclei that have this property include ^{1}H, ^{13}C and ^{19}F.

Examiners' tip

It is not necessary to understand the science behind the resonance processes involved in n.m.r. — it can be highly complex. But you must be able to appreciate the importance of various aspects to interpret an n.m.r. spectrum.

How to understand an n.m.r. spectrum

Revised

The following aspects are important to understand when interpreting an n.m.r. spectrum.

- A compound called tetramethylsilane (TMS), $Si(CH_3)_4$, (see Figure 9.3) is used as a reference to which other proton resonances are compared. It is used to calibrate the signals produced when analysing the compound under test.

  Figure 9.3 Tetramethylsilane, TMS

  TMS has four methyl groups — all its protons are the same, or **equivalent**, and therefore resonate at the same frequency in a magnetic field.

  A few drops of TMS are all that is required because it produces a strong signal, due to its 12 protons all resonating at the same frequency. TMS can be easily removed from the sample under test because it is volatile and chemically inert.
- The **chemical shift** (δ) axis on the spectrum produced measures the resonances at which protons in the sample occur in relation to the resonance of the protons in TMS reference. It is measured in parts per million (ppm).
- The further to the left-hand side a peak occurs along the chemical shift axis (with higher δ values), the closer the protons (or ^{13}C atoms) are to electronegative groups, like oxygen, in the molecule.
- The chemical shift tells us about the **chemical environment** of the protons (or ^{13}C atoms) in the molecule.
- The vertical axis measures the absorptions of energy that are happening.
- The **areas under the peaks** in a proton n.m.r. spectrum give us information about the number of hydrogen atoms giving rise to the absorptions.

There are two types of n.m.r. to cover — ^{13}C n.m.r. and ^{1}H (or proton) n.m.r.

^{13}C spectra

The chemical shifts for ^{13}C atoms together with the corresponding groups giving rise to the absorptions are shown in Figure 9.4.

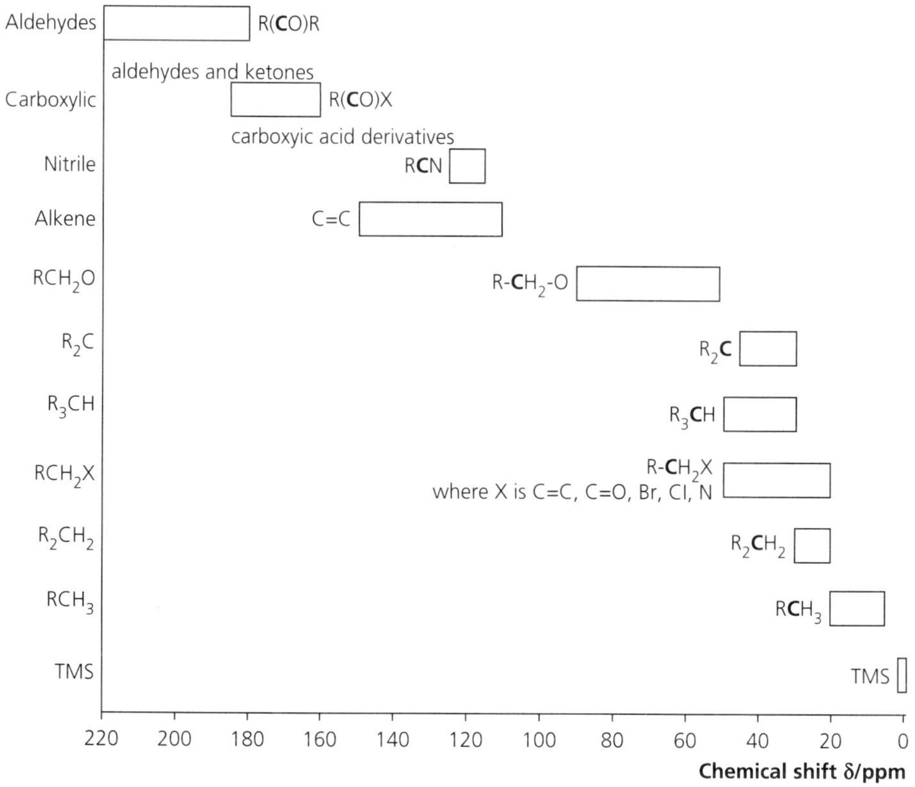

Figure 9.4 Chemical shifts

Figure 9.5 shows the ^{13}C spectrum for ethyl ethanoate.

$$CH_3\overset{\displaystyle O}{\overset{\|}{C}}OCH_2CH_3$$

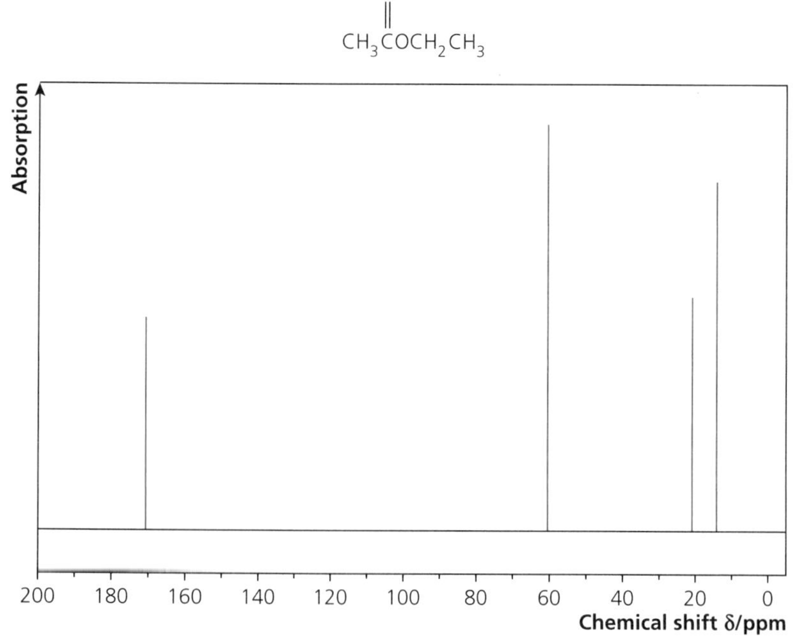

Figure 9.5 A typical ^{13}C n.m.r. spectrum

Note the following from Figure 9.5:

- The number of peaks tell us the number of different carbon atom environments in the molecule under test — four in this case.
- The peak at $\delta = 170$ is due to a carbonyl carbon because it is attached directly to an electronegative oxygen atom.

Examiners' tip

^{13}C n.m.r. spectroscopy is a lot more straightforward to understand than ^1H spectroscopy. It is often just a matter of counting the number of peaks (this gives the number of different types of carbon atom) and reading the positions of each peak along the chemical shift axis (this gives the chemical environment).

- The peak at $\delta = 60$ is due to a CH_2 carbon attached to the oxygen of the acyl group of the ester.
- The peaks at $\delta = 20$ and 15 are due to two CH_3 groups.

^{1}H spectra

Revised

A ^{1}H n.m.r. spectrum is similar to a ^{13}C n.m.r. spectrum but the areas under the peaks and spin–spin coupling are extra considerations that make it a very powerful technique for deducing the identity of a molecule.

Spin–spin coupling happens when the spin of one proton couples with the spins of neighbouring **non-equivalent** protons and causes a signal to split.

Table 9.2 lists chemical shifts that will be of use in the questions that follow this section.

Table 9.2 Chemical shifts in ^{1}H n.m.r.

Type of proton	δ/ppm
RO**H**	0.5–5.0
RC**H**$_3$	0.7–1.2
RN**H**$_2$	1.0–4.5
R_2C**H**$_2$	1.2–1.4
R_3C**H**	1.4–1.6
R—C(=O)—C(**H**)—	2.1–2.6
R—O—C(**H**)—	3.1–3.9
RC**H**$_2$Cl or Br	3.1–4.2
R—C(=O)—O—C(**H**)—	3.7–4.1
R(—)C=C(**H**)(—)	4.5–6.0
R—C(=O)—**H**	9.0–10.0
R—C(=O)—O—**H**	10.0–12.0

Figure 9.6 shows the ^{1}H n.m.r. spectrum for ethanol, C_2H_5OH

```
    H   H
    |   |
H — C — C — O — H
    |   |
    H   H
```

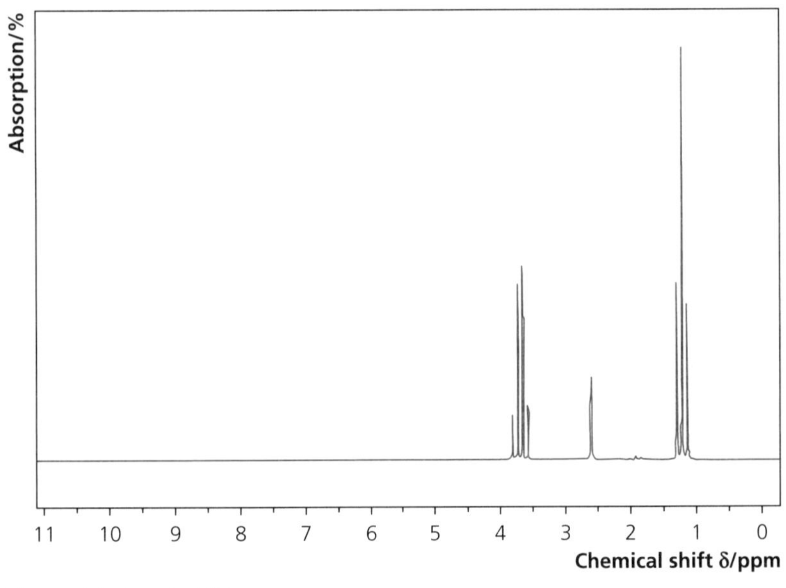

Figure 9.6 A typical ^{1}H n.m.r. spectrum

Important features of this spectrum include:

- 3 main peaks, so 3 proton environments
- a triplet at $\delta = 1.2$ due to the CH_3 protons
- a singlet at $\delta = 2.6$ due to the O–H proton
- a quartet at $\delta = 3.7$ due to the CH_2 protons.

The areas under the peaks in Figure 9.6, reading from left to right, will have a ratio 2:1:3 because these are in proportion to the number of hydrogen atoms in each proton environment giving rise to each absorption.

Figure 9.6 also shows the displayed formula of ethanol.

- The CH_3 protons are next to the two CH_2 protons, and the CH_3 resonance is split by these **neighbouring protons**. The amount of splitting depends on the number of hydrogens 'next door' + 1, or $n + 1$. So, because there are two protons on the carbon atom next to the CH_3 group, the CH_3 resonance is split into 2 + 1 = 3, or a triplet.

Examiners' tip

Remember the '$n + 1$' rule — the number of splits in an absorption is equal to the number of neighbouring protons + 1.

- The pattern of these three split peaks can be judged from Pascal's triangle:

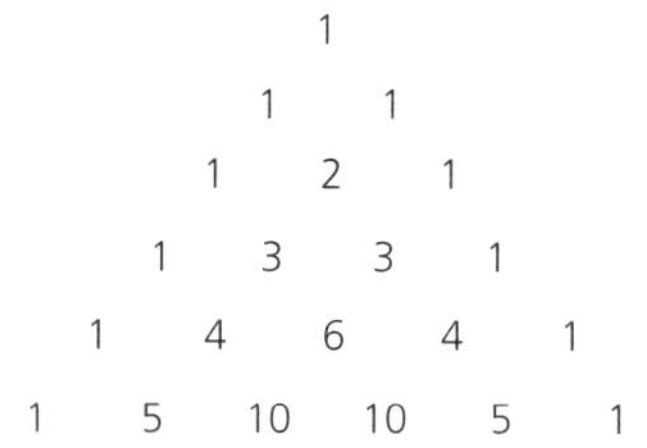

Three numbers in a horizontal row correspond to 1:2:1. So a **1:2:1 triplet** is observed.

- The CH_2 protons are next to CH_3 protons, and so the CH_2 resonance will be split into 3 + 1 = 4, a quartet is observed. The pattern of this quartet will be a **1:3:3:1 quartet**.
- The OH proton absorption is not split and this is the case in general — a proton attached to an electronegative atom like oxygen, does not experience any splitting from neighbouring protons. It therefore appears as a **singlet**.

To conclude, the proton environments indicated in Figure 9.7 have now been associated with the adsorptions shown.

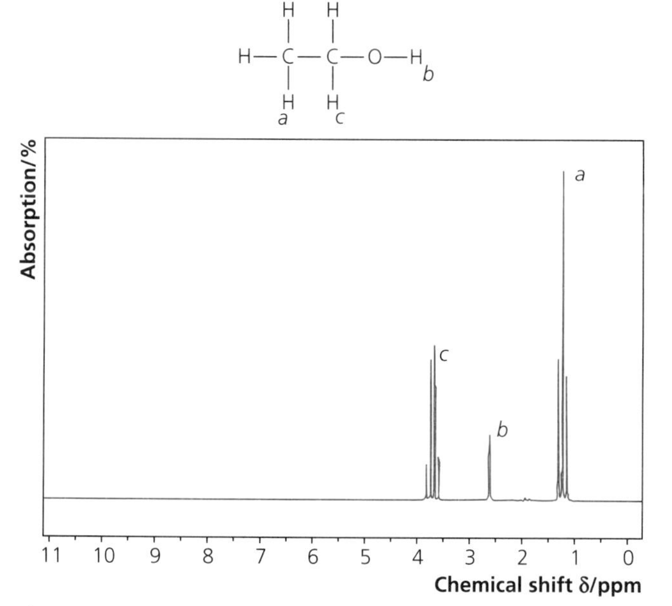

Figure 9.7 ^{1}H n.m.r. spectrum of ethanol

Now test yourself

2 Consider the molecule shown by its displayed formula:

Predict and explain what its proton n.m.r. spectrum would look like.

Answers on p. 110

Tested

Examiners' tip

When considering ^{1}H and ^{13}C n.m.r. spectra in examination questions, remember to mention the number of peaks and the chemical shifts. In ^{1}H spectra you will also need to consider areas under peaks and splitting patterns.

Chromatography

Gas–liquid chromatography can be used to separate mixtures of volatile liquids. Figure 9.8 shows a schematic diagram of the apparatus used.

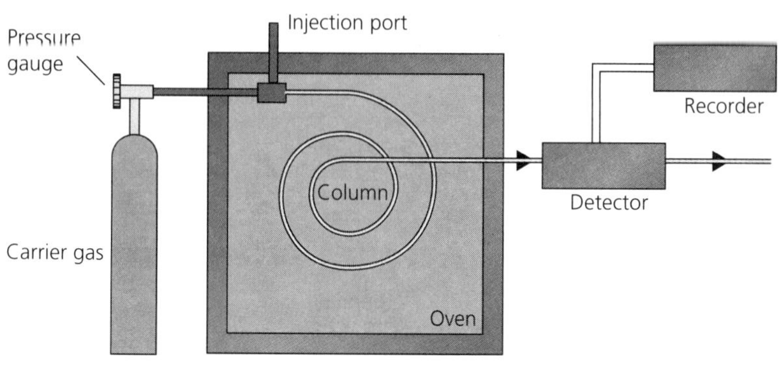

Figure 9.8 Column chromatography

The sample under test is injected into the column and heated. The oven contains of a long, narrow copper tube in which a material — for example, aluminium oxide — acts as the **stationary phase**. Many substances can be used as the stationary phase — some are polar, like aluminium oxide, and others are non-polar, like alkane-based materials. The sample mixture is moved through the stationary phase in the copper tube using an inert gas under pressure (**mobile phase**). The sample is separated into its individual components according to their **solubility** in the mobile phase and their **retention** in the stationary phase.

Examiners' tip

The length of time a compound spends in the column depends on the solubility in the stationary phase, and also how it is affected by the mobile phase. A polar species will spend a long time in the column if a polar stationary phase is used.

One substance that can be analysed using gas–liquid chromatography is premium grade petrol, see Figure 9.9 for a typical chromatogram.

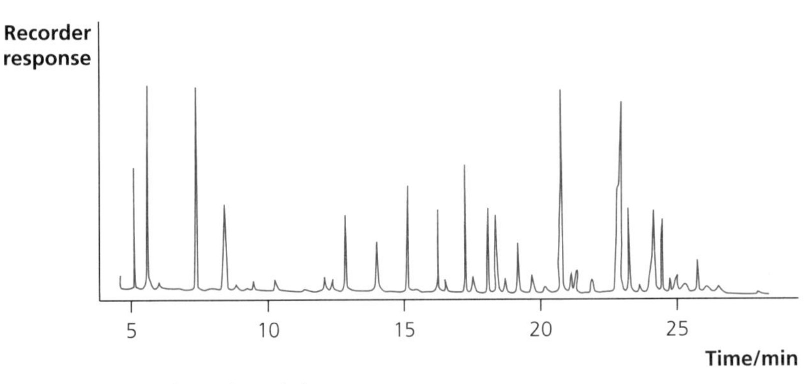

Figure 9.9 Typical gas–liquid chromatogram

Notice that there are many peaks, each representing a different compound that comes out of the chromatography column. Even after a short retention time of a few minutes, compounds start to emerge from the column — these are relatively insoluble in the stationary phase but soluble in the mobile phase. Those that are retained in the column for 20–25 minutes, are considerably more soluble in the stationary phase and take more time to be removed.

Exam practice

1 An ester, Y, is analysed to produce the three spectra shown below.

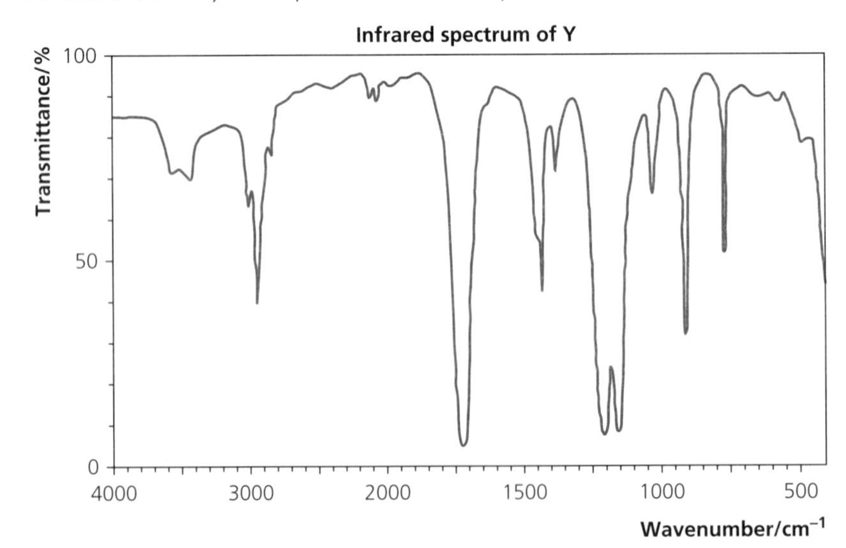

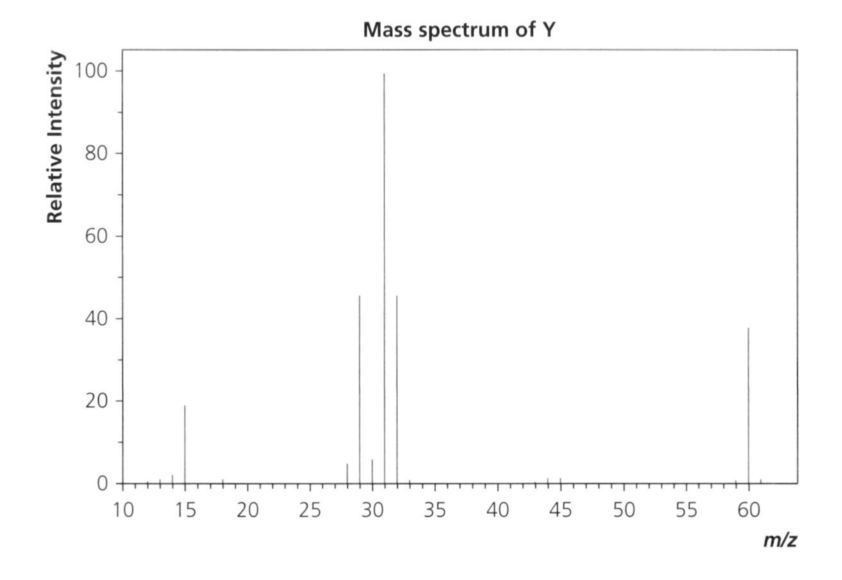

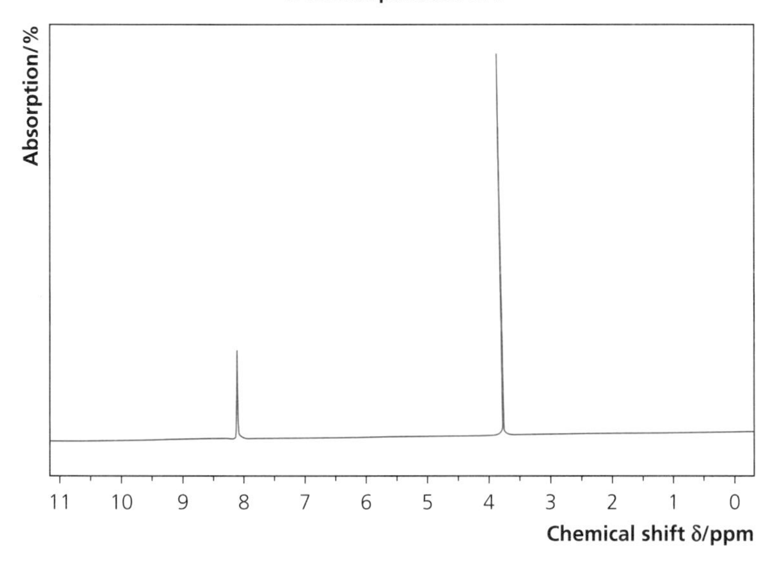

(a) Using the infrared spectrum of Y, confirm that Y may be an ester. [2]

(b) Using the mass spectrum of Y, deduce its relative molecular mass. [1]

(c) Suggest the formulae of the fragments responsible for the peaks at $m/z = 31$ and $m/z = 29$ in the mass spectrum of Y. [2]

(d) Using the ^{1}H n.m.r. spectrum for compound Y, suggest:

(i) how many proton environments are present in Y [1]

(ii) the nature of these environments. [1]

(e) Draw the displayed formula for Y and name the ester. [2]

2 A student carries out an experiment in which a compound with the formula $NaBH_4$ is added to ethanal. He obtains two spectra from the starting material — a mass spectrum and a ^{1}H n.m.r. spectrum; and two spectra from the product of the reaction — an infrared spectrum and a mass spectrum.

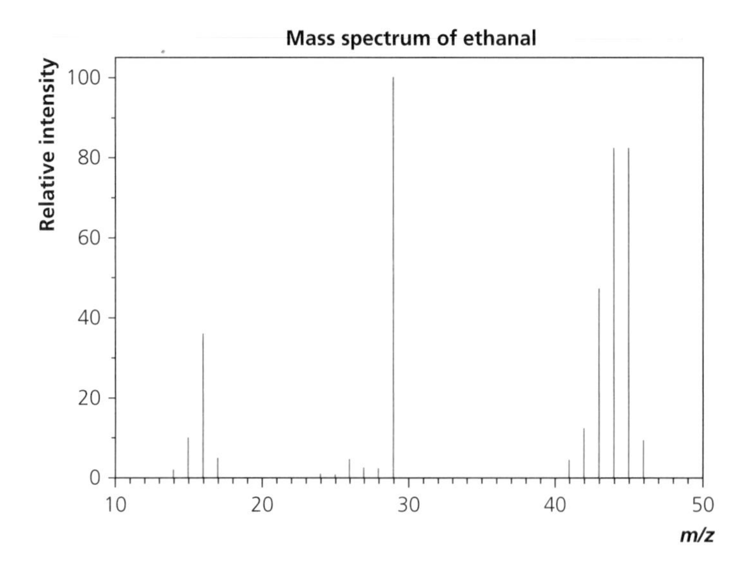
Mass spectrum of ethanal
Relative intensity
100
80
60
40
20
0
10
20
30
40
50
m/z

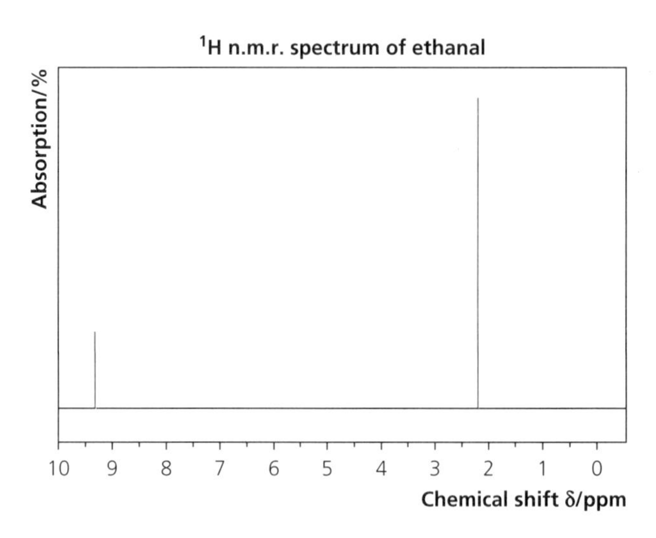
1H n.m.r. spectrum of ethanal
Absorption/%
10
9
8
7
6
5
4
3
2
1
0
Chemical shift δ/ppm

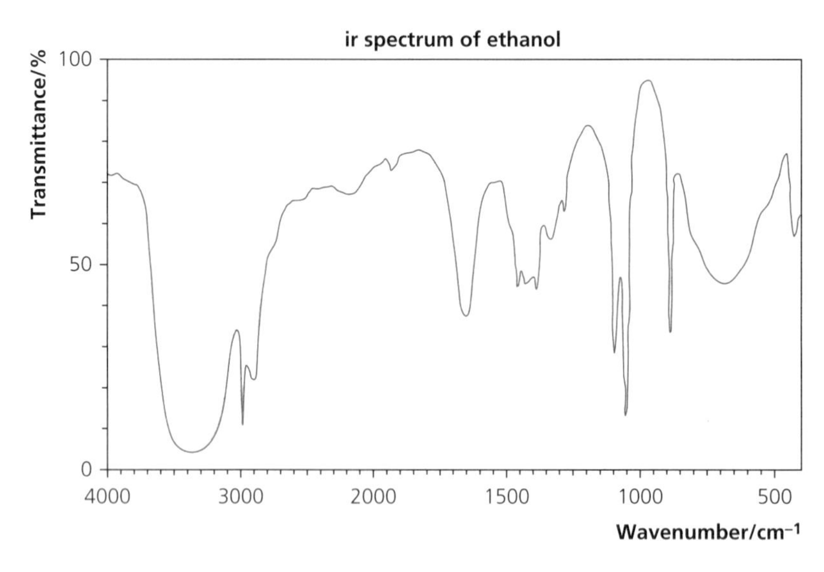
ir spectrum of ethanol
Transmittance/%
100
50
0
4000
3000
2000
1500
1000
500
Wavenumber/cm−1

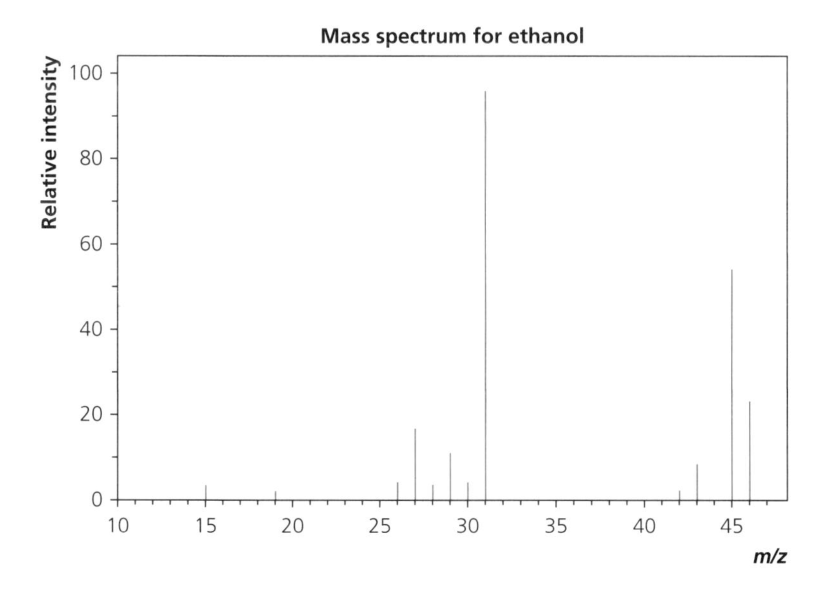

(a) Use the mass spectrum for ethanal to deduce its relative molecular mass. Explain how you arrived at your answer. [2]

(b) What is the species giving rise to the peak at $m/z = 29$? [1]

(c) Identify the proton environments giving rise to the absorptions in ethanal's n.m.r. spectrum. [2]

(d) Explain how ethanol's infrared spectrum indicates that an alcohol has been produced. [1]

(e) A peak occurs at $m/z = 31$ in ethanol's mass spectrum. Explain the existence of this peak. [1]

In a further reaction, ethanal reacts with hydrogen cyanide/KCN and an absorption is observed at $3405\,cm^{-1}$ in the infrared spectrum of the product shown below.

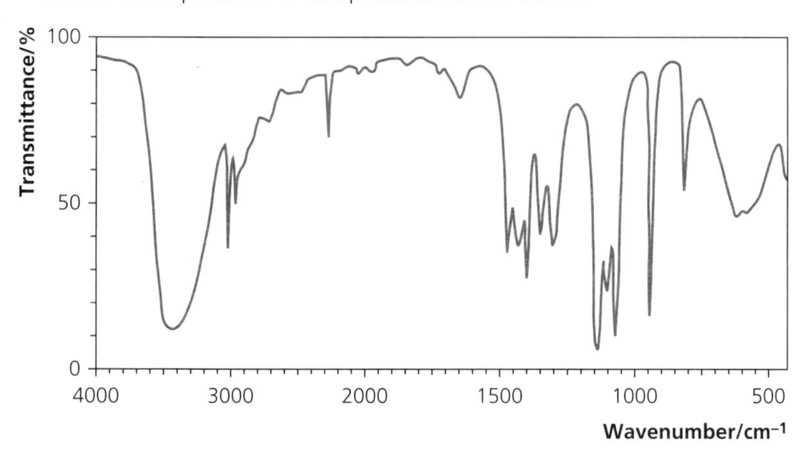

(f) Explain the existence of the peak at $3405\,cm^{-1}$ with reference to the likely product. [2]

Answers and quick quizzes online

Online

Examiners' summary

You should now have an understanding of:

- ✔ how mass spectroscopy can be used to identify compounds
- ✔ the fact that molecules can undergo fragmentation and that some of the resulting positive ions are more stable than others
- ✔ how infrared spectroscopy is used to identify specific bonds in molecules
- ✔ the information that ^{13}C n.m.r. spectra can provide
- ✔ how proton n.m.r spectra is interpreted to yield valuable information about the structure of compounds
- ✔ why TMS is used in n.m.r. spectroscopy
- ✔ the 'n + 1 rule' and how it is used to interpret spectra
- ✔ how gas–liquid chromatography and column chromatography can be used to separate mixtures in terms of the polar nature of components and the phases used

10 Thermodynamics

Enthalpy change

There are some important definitions that you must learn and also apply, some of which you will already know from the AS level course.

All the energy terms listed below are considered to be **standard values** — that is their measurement is done at 298 K and a pressure of 100 kPa.

- An **enthalpy of formation** is the enthalpy change when 1 mole of a substance is formed from its constituent elements, with all reactants and products in their standard states. For example, using calcium carbonate:
  $Ca(s) + C(s) + 1\frac{1}{2}O_2(g) \rightarrow CaCO_3(s)$
- A **first ionisation enthalpy** is the enthalpy change for the removal of 1 mole of electrons from 1 mole of gaseous atoms to produce 1 mole of singly charged positive ions in the gas phase. For example, using magnesium:
  $Mg(g) \rightarrow Mg^+(g) + e^-$
- An **enthalpy of atomisation** is the enthalpy change for the formation of 1 mole of gaseous atoms from an element in its standard state. For example, using oxygen:
  $\frac{1}{2}O_2(g) \rightarrow O(g)$
- A **bond dissociation enthalpy** is the enthalpy change required to break 1 mole of a specific bond in a specific compound in the gas phase. For example, using ammonia:
  $NH_3(g) \rightarrow NH_2(g) + H(g)$
- A **mean bond dissociation enthalpy** is the average of the bond dissociation energies for a given bond measured over many compounds.
- An **electron affinity** is the enthalpy change when 1 mole of electrons is added to 1 mole of gaseous atoms to form 1 mole of gaseous negative ions. For example, using chlorine:
  $Cl(g) + e^- \rightarrow Cl^-(g)$
- A **lattice formation enthalpy** is the enthalpy change when 1 mole of an ionic solid is formed from its constituent gaseous ions. For example, using magnesium chloride:
  $Mg^{2+}(g) + 2Cl^-(g) \rightarrow MgCl_2(s)$
- A **lattice dissociation enthalpy** is the enthalpy change when 1 mole of an ionic solid is turned into its constituent gaseous ions. For example, using magnesium chloride:
  $MgCl_2(s) \rightarrow Mg^{2+}(g) + 2Cl^-(g)$
- An **enthalpy of hydration** is the enthalpy change when 1 mole of a gaseous ion forms in a hydrated ion. For example, using lithium ions:
  $Li^+(g) + aq \rightarrow Li^+(aq)$

Examiners' tip

Learn these definitions — they are often requested in examinations — but also understand their chemical significance. There is often a change of physical state and this can make a huge difference, so do take notice of any state symbols.

Typical mistake

Most candidates realise that most of the definitions on this page are defined according to *starting* with 1 mole of substance. However, enthalpy of atomisation is defined according to the *formation* of 1 mole of gaseous atoms. Be careful.

- An **enthalpy of solution** is the enthalpy change when 1 mole of a substance is dissolved in water to infinite dilution. For example, using sodium sulfate(VI):

  $Na_2SO_4(s) \rightarrow 2Na^+(aq) + SO_4^{2-}(aq)$

The Born–Haber cycle

Revised

A lattice enthalpy can be analysed or calculated using a **Born–Haber cycle**.

Example

Draw a Born–Haber cycle for the formation of calcium fluoride, CaF_2, and use it to calculate the lattice formation enthalpy.

Answer

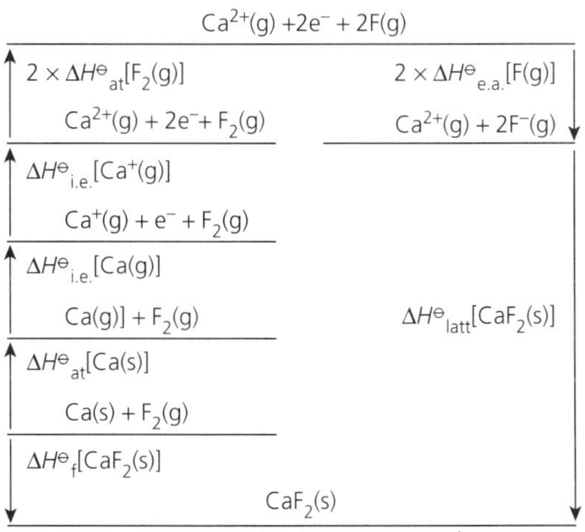

$Ca^{2+}(g) + 2e^- + 2F(g)$

$2 \times \Delta H^\ominus_{at}[F_2(g)]$ ↑ $2 \times \Delta H^\ominus_{e.a.}[F(g)]$ ↓

$Ca^{2+}(g) + 2e^- + F_2(g)$ $Ca^{2+}(g) + 2F^-(g)$

$\Delta H^\ominus_{i.e.}[Ca^+(g)]$ ↑

$Ca^+(g) + e^- + F_2(g)$

$\Delta H^\ominus_{i.e.}[Ca(g)]$ ↑

$Ca(g)] + F_2(g)$ $\Delta H^\ominus_{latt}[CaF_2(s)]$ ↓

$\Delta H^\ominus_{at}[Ca(s)]$ ↑

$Ca(s) + F_2(g)$

$\Delta H^\ominus_f[CaF_2(s)]$ ↓

$CaF_2(s)$

Relevant standard enthalpy data (in $kJ\,mol^{-1}$):

$\Delta H^\ominus_f[CaF_2(s)] = -1214$
$\Delta H^\ominus_{at}[Ca(s)] = +193$
$\Delta H^\ominus_{i.e.}[Ca(g)] = +590$
$\Delta H^\ominus_{i.e.}[Ca^+(g)] = +1150$
$\Delta H^\ominus_{at}[F_2(g)] = +79$
$\Delta H^\ominus_{e.a}[F(g)] = -348$

Figure 10.1 Born–Haber cycle for CaF_2

Using the cycle:

$\Delta H^\ominus_{at}[Ca(s)] + \Delta H^\ominus_{i.e.}[Ca(g)] + \Delta H^\ominus_{i.e.}[Ca^+(g)] + 2 \times \Delta H^\ominus_{at}[F_2(g)] + 2 \times \Delta H^\ominus_{e.a.}[F(g)] + \Delta H^\ominus_{latt}[CaF_2(s)] = \Delta H^\ominus_f[CaF_2(s)]$

$\Delta H^\ominus_{latt}[CaF_2(s)] = \Delta H^\ominus_f[CaF_2(s)] - \Delta H^\ominus_{at}[Ca(s)] - \Delta H^\ominus_{i.e.}[Ca(g)] - \Delta H^\ominus_{i.e.}[Ca^+(g)] - 2 \times \Delta H^\ominus_{at}[F_2(g)] - 2 \times \Delta H^\ominus_{e.a.}[F(g)]$

$= (-1214) - (+193) - (+590) - (+1150) - (2 \times +79) - (2 \times -348)$

$= -1214 - 193 - 590 - 1150 - 158 + 696$

$= -2609\,kJ\,mol^{-1}$

Notice that the lattice energy is **highly exothermic**. This is because attractions are being formed between the gaseous ions as they come together to form their ionic lattice.

It is likely that questions will be set in examinations in which you are asked to calculate any one of the terms in the Born–Haber cycle given all others.

lattice enthalpy

thalpy is related to the **charges** on the ions and their **ionic** he higher the ionic charges and the smaller the ionic radii, the more exothermic will be the lattice formation enthalpy. For example, sodium fluoride has a value for $\Delta H^{\ominus}_{latt}$ of $-902\,kJ\,mol^{-1}$, whereas magnesium oxide has a value of $-3889\,kJ\,mol^{-1}$. The Mg^{2+} ion is smaller than the Na^{+} ion, and it also has a higher charge; the O^{2-} ion is a similar size to the F^{-} ion but it has a higher negative charge. Therefore, the Mg^{2+} and O^{2-} ions will be more strongly attracted in MgO than the Na^{+} and F^{-} ions in NaF. MgO therefore has a more exothermic lattice formation enthalpy.

It is possible to calculate a **theoretical value** for the lattice formation enthalpy of an ionic solid — this assumes that the solid is 100% ionic in character. When such values are compared with those obtained using a Born–Haber cycle (for which the values are obtained experimentally), evidence can be gained about the type of bonding present in a compound.

If there is a significant difference between the 100% ionic model and the experimental Born–Haber cycle value (see Table 10.1) then we can assume that there must be some **covalent character** in the substance. This is attributed to the **polarising power** of small and highly charged positive ions distorting the spherical electron distribution of the negative ions thereby **inducing** some covalent character and strengthening the lattice.

Table 10.1 Covalent character in ionic compounds

Substance	Theoretical lattice enthalpy/$kJ\,mol^{-1}$	B–H lattice enthalpy/ $kJ\,mol^{-1}$	Difference/%
LiCl	833	846	1.56
NaCl	766	771	0.65
KCl	690	701	1.59
RbCl	674	675	0.15
AgCl	770	905	17.5
AgBr	758	890	18.7
AgI	736	876	19.0

The following deductions can be made:

- The ionic model fits better at the bottom of group 1 than the top. Lithium ions are smaller than the other ions in the group, and so they will have a bigger polarising power towards the chloride ions, hence inducing more covalent character in the compound.
- There is more **ionic character** in the compounds as group 1 is descended.
- There is, conversely, more covalent character moving up the group.
- Differences are much more marked in other regions of the periodic table. The percentage difference in the silver halide values is over ten times greater than the difference in the group 1 halides. Therefore silver halides have an appreciable covalent character which is reflected by their low soluability in water.

Lattice enthalpies can also be used for calculating the **enthalpy of solution** of a compound. An energy cycle is used that combines the lattice enthalpy of the compound and the hydration enthalpies of the ions with the enthalpy of solution:

Examiners' tip

When calculating an enthalpy change using a cycle such as the Born–Haber cycle, apply Hess's law to find an alternative route. Remember to change the sign of any term that points in the opposite direction to the one you want to move along.

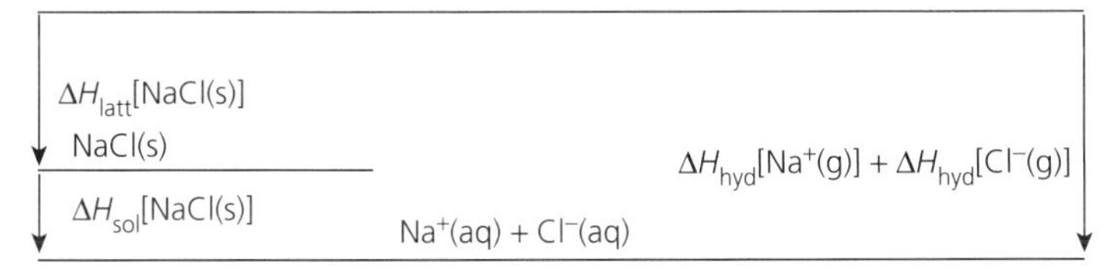

Figure 10.2 Calculating enthalpy of solution

$\Delta H^\ominus_{latt}[NaCl(s)] = -771\,kJ\,mol^{-1}$; $\Delta H^\ominus_{hyd}[Na^+(g)] = -406\,kJ\,mol^{-1}$;
$\Delta H^\ominus_{hyd}[Cl^-(g)] = -364\,kJ\,mol^{-1}$

It can be seen from the cycle that:

$$\Delta H^\ominus_{sol}[NaCl(s)] = \Delta H^\ominus_{hyd}[Na^+(g)] + \Delta H^\ominus_{hyd}[Cl^-(g)] - \Delta H^\ominus_{latt}[NaCl(s)]$$

If the sum of the hydration enthalpies is greater than the lattice enthalpy, then the substance will have a negative value for the enthalpy of solution, and vice versa.

For sodium chloride:

$$\Delta H^\ominus_{sol}[NaCl(s)] = (-406) + (-364) - (-771)$$
$$= -406 - 364 + 771$$
$$= +1\,kJ\,mol^{-1}$$

The enthalpy of solution for sodium chloride is therefore only slightly endothermic — this is because the lattice enthalpy is balanced by the sum of the hydration enthalpies almost exactly.

Examiners' tip

Lattice enthalpy and hydration enthalpies depend on the same factors — the higher the ionic charges and the smaller the ionic radii, the greater will be the attraction between the ions in the lattice, *and* the attraction between water molecules and the individual ions.

Now test yourself — Tested

The ionic radii, in nm, of some positive and negative ions are given in the table. Use the data to answer the questions that follow.

Positive ions	Ionic radii/nm	Negative ions	Ionic radii/nm
K^+	0.133	F^-	0.136
Ca^{2+}	0.100	Cl^-	0.188
		Br^-	0.195
		S^{2-}	0.185

1 Predict which of the following compounds will have the *largest* difference between its lattice energy calculated from experimental values and theoretical value calculated assuming 100% ionic character.
A CaS **B** K_2O **C** KF **D** CaO

2 Predict which of the following will have the most exothermic lattice formation enthalpy. Assume that they all have the same crystal structure.
A CaS **B** CaO **C** KF **D** KCl

3 Predict which of the ions below is likely to have the most exothermic hydration enthalpy.
A F^- **B** Ca^{2+} **C** Br^- **D** K^+

Answers on p. 110

...enthalpies

Revised

...**ssociation enthalpy** (b.e.) is the enthalpy change required to ...nole of a specific bond in a specific compound in the gas phase.

There is often a big difference between the bond strength in a particular compound and the average value worked out over many compounds.

Nearly always, in calculations, except for elements, it is the mean values that are quoted (in kJ mol^{-1}) — for example:

N–H = 388; H–Cl = 431; O–H = 463; N–N = 145

H–H = 436; O–O = 157; O=O = 496; Cl–Cl = 242; N≡N = 944

Example 1

Calculate the enthalpy change for the reaction that takes place between hydrazine and hydrogen peroxide:

$N_2H_4(g) + 2H_2O_2(g) \rightarrow N_2(g) + 4H_2O(g)$

In terms of bonds, this looks like:

Answer

Using Hess's law, we can draw this energy cycle:

$N_2H_4(g) + 2H_2O_2(g) \xrightarrow{\Delta H_1} N_2(g) + 4H_2O(g)$

ΔH_2: $N_2H_4(g) + 2H_2O_2(g) \rightarrow 2N(g) + 8H(g) + 4O(g)$

ΔH_3: $N_2(g) + 4H_2O(g) \rightarrow 2N(g) + 8H(g) + 4O(g)$

It can be seen that $\Delta H_1 = \Delta H_2 - \Delta H_3$

Using the bond energies above:

ΔH_2 = b.e.(N–N) + 4 × b.e.(N–H) + 2 × b.e.(O–O) + 4 × b.e.(O–H)

ΔH_3 = b.e.(N≡N) + 8 × b.e.(O–H)

So ΔH_1 = (145) + (4 × 388) + (2 × 157) + (4 × 463) – (944) – (8 × 463)

= 145 + 1552 + 314 + 1852 – 944 – 3704

= –785 kJ mol^{-1}

Example 2

Calculate the enthalpy change for the reaction below in which substances are all in their standard states:

$N_2H_4(l) + 2H_2O_2(l) \rightarrow N_2(g) + 4H_2O(l)$

Enthalpy of formation data:

$\Delta H^\ominus_f$/kJ mol^{-1}: $N_2H_4(l) = +50$; $H_2O_2(l) = -188$; $H_2O(l) = -286$

Answer

From the AS course, $\Delta H = \Sigma\Delta H_f(\text{products}) - \Sigma\Delta H_f(\text{reactants})$

$$= [(4 \times -286) + 0] - [(+50) + (2 \times -188)]$$

$$= -1144 + 326$$

$$= -818\text{ kJ mol}^{-1}$$

Comparing this value with that obtained in example 1, it can be seen that there is a difference of 33 kJ mol^{-1}. The reasons for this include:

- In example 1, mean bond enthalpies were being used for compound bonds when specific bond enthalpies for the substances in the reaction would have been more accurate.
- In example 1, hydrogen peroxide, hydrazine and water were all in the gas phase. Their standard state is liquid.

Now test yourself Tested

4 Calculate ΔH for the reactions below using bond enthalpies given in this section.
 (a) $N_2(g) + 2H_2(g) \rightarrow N_2H_4(g)$
 (b) $H_2O_2(g) \rightarrow H_2(g) + O_2(g)$
 (c) $H_2O_2(g) \rightarrow H_2O(g) + \frac{1}{2}O_2(g)$

5 The enthalpy of formation of hydrazine, from:
 $N_2(g) + 2H_2(g) \rightarrow N_2H_4(l)$
 is +50.4 kJ mol^{-1}
 Comment on any differences between this value and that from question 4(a).

Answers on p. 110

Entropy and Gibb's free energy

Entropy, S, is a measure of the amount of **disorder** of the particles being considered. For example, the standard entropies ($S^\ominus$), measured at 298 K, of three elements are shown in Table 10.2.

The **entropies** of the three main physical states increase in the order solid to liquid to gas.

It can be seen that the values increase on moving from a solid to a liquid to a gas. This is because there is more disorder in a gaseous system and there will, therefore, be a greater number of ways that the energy possessed by the molecules present can be arranged.

Table 10.2 Entropy values for some elements

Element	$S^\ominus$/J K^{-1} mol^{-1}
Fe(s)	27.2
Hg(l)	77.4
O_2(g)	205

Many processes have associated with them a spontaneous **increase of disorder**, or increase in entropy — for example:

- **Melting**

  $H_2O(s) \rightarrow H_2O(l)$
- **Evaporation**

  $C_2H_5OH(l) \rightarrow C_2H_5OH(g)$
- **Dissolution** or dissolving

  $CaCl_2(s) + (aq) \rightarrow Ca^{2+}(aq) + 2Cl^-(aq)$
- **Formation of a gas**

  $NaHCO_3(s) + HCl(aq) \rightarrow NaCl(aq) + H_2O(l) + CO_2(g)$

Now test yourself Tested

6 State whether each of the following processes involves an increase or decrease in entropy.

(a) $CuSO_4(s) + 5H_2O(l) \rightarrow CuSO_4.5H_2O(s)$

(b) $Mg(s) + H_2SO_4(aq) \rightarrow MgSO_4(aq) + H_2(g)$

(c) $NaCl(s) + (aq) \rightarrow NaCl(aq)$

(d) $2CO(g) + O_2(g) \rightarrow 2CO_2(g)$

(e) $H_2O(l) \rightarrow H_2O(s)$

Answers on p. 110

Calculating entropy changes

When a process happens, it is possible to calculate the entropy change by working out the difference between the final entropy and the initial entropy. For example, under standard conditions:

$$\Delta S^{\ominus} = \Sigma S^{\ominus}(\text{products}) - \Sigma S^{\ominus}(\text{reactants})$$

Example

Calculate the standard entropy change for the following process given the individual absolute standard entropy values:

	$2NaHCO_3(s)$	$\rightarrow Na_2CO_3(s)$	$+ H_2O(l)$	$+ CO_2(g)$
$S^{\ominus}/J\,K^{-1}\,mol^{-1}$	102	136	70	214

$\Delta S^{\ominus} = \Sigma S^{\ominus}(\text{products}) - \Sigma S^{\ominus}(\text{reactants})$

$\Delta S^{\ominus} = [136 + 70 + 214] - [2 \times 102]$

$= 420 - 204$

$= +216\,J\,K^{-1}\,mol^{-1}$

The entropy change is highly positive — this was predictable because a gas was formed. An increase in the number of moles of gas always leads to an increase in entropy.

Typical mistake

In questions like this, many candidates forget to multiply the $NaHCO_3$ entropy by 2.

Now test yourself Tested

7 **(a)** Calculate the standard entropy change, $\Delta S^{\ominus}$, for the reaction:

	$CaO(s)$	$+ CO_2(g)$	$\rightarrow CaCO_3(s)$
$S^{\ominus}/J\,K^{-1}\,mol^{-1}$	40	214	93

(b) Comment on the sign of the entropy value and explain why this should be expected.

Answers on p. 110

Gibbs free energy, ΔG

Revised

Enthalpy change alone cannot be used to predict whether or not a reaction will take place on its own. Reactions that are spontaneous are also described as **feasible**.

ΔG must be negative for a reaction to occur.

Although many reactions that take place readily are exothermic, there are also examples of endothermic reactions that are spontaneous — for example the dissolving of barium nitrate, $Ba(NO_3)_2(s)$, in water:

$$Ba(NO_3)_2(s) + aq \rightarrow Ba^{2+}(aq) + 2NO_3^-(aq); \Delta H = +40.4\,kJ\,mol^{-1}$$

Clearly, a better way of determining whether reactions are spontaneous or not is required. This uses the **Gibb's free energy change**, ΔG, which is the true indicator of the expected direction of chemical change.

A Gibb's free energy change is calculated using the relationship:

$$\Delta G = \Delta H - T\Delta S$$

- ΔG is the Gibb's free energy change in $kJ\,mol^{-1}$
- ΔH is the molar enthalpy change for the reaction in $kJ\,mol^{-1}$
- T is the Kelvin temperature
- ΔS is the entropy change in $kJ\,K^{-1}\,mol^{-1}$

Example

Given the data provided, calculate $\Delta G^\ominus$, at 298 K, for the reaction:

	ZnO(s) +	CO(g) →	Zn(s) +	$CO_2(g)$
$S^\ominus/J\,K^{-1}\,mol^{-1}$	44	198	42	214
$\Delta H^\ominus_f/kJ\,mol^{-1}$	−348	−111	0	−394

Answer

Calculate the standard entropy change for the reaction using $\Delta S^\ominus = \Sigma S^\ominus(\text{products}) - \Sigma S^\ominus(\text{reactants})$:

$$\Delta S^\ominus = (42 + 214) - (44 + 198)$$
$$= 256 - 242$$
$$= +14\,J\,K^{-1}\,mol^{-1} = 0.014\,kJ\,K^{-1}\,mol^{-1}$$

Calculate the enthalpy change for the reaction using $\Delta H^\ominus = \Sigma H^\ominus(\text{products}) - \Sigma H^\ominus(\text{reactants})$:

$$\Delta H^\ominus = (-394 + 0) - (-348 + -111)$$
$$= -394 - (-459)$$
$$= -394 + 459$$
$$= +65\,kJ\,mol^{-1}$$

Calculate the standard Gibb's free energy change using $\Delta G^\ominus = \Delta H^\ominus - T\Delta S^\ominus$:

$$\Delta G^\ominus = 65 - (298 \times 0.014)$$
$$= 65 - 4.2$$
$$= +60.8\,kJ\,mol^{-1}$$

This reaction is not spontaneous at 298 K because the sign for $\Delta G^\ominus$ is positive. So, it is not possible for carbon monoxide to reduce zinc oxide to form zinc and carbon dioxide at 298 K.

Typical mistake

Many candidates forget to divide the entropy term by 1000 to convert from J to kJ when they use $\Delta G = \Delta H - T\Delta S$.

Deducing the temperature at which a reaction becomes spontaneous

Another term for 'spontaneous' is 'feasible'.

Looking at the equation $\Delta G = \Delta H - T\Delta S$, it can be seen that:

- as the temperature increases, $T\Delta S$ becomes more positive
- as $T\Delta S$ becomes more positive, it must reach a temperature at which $T\Delta S$ exceeds ΔH
- at this temperature, ΔG will start to be negative and the reaction is now spontaneous.

Example

Using this reaction again:

$$ZnO(s) + CO(g) \rightarrow Zn(s) + CO_2(g)$$

at what temperature does it become feasible?

At the temperature at which ΔG changes from being positive to negative, $\Delta G = 0$ or when $\Delta H = T\Delta S$

So $T = \frac{\Delta H}{\Delta S}$

But $\Delta H = +65\,kJ\,mol^{-1}$ and $\Delta S = +0.014\,kJ\,K^{-1}\,mol^{-1}$

$T = \frac{65}{0.014}$

$= 4643\,K$ or $4370°C$

This means that carbon monoxide can reduce zinc oxide at or above 4370°C. Such a high temperature is very expensive to produce so the process is probably an uneconomic method for producing zinc.

A closer look at $\Delta G = \Delta H - T\Delta S$

We can apply the equation to two general cases.

Reactions with a **negative entropy change**.

For example, $N_2(g) + 3H_2(g) \rightarrow 2NH_3(g)$:

- ΔS is negative, and so $T\Delta S$ will also be negative
- As temperature increases, $T\Delta S$ becomes more negative
- $\Delta H - T\Delta S$ therefore becomes more positive as temperature increases
- ΔG therefore becomes more positive
- So a temperature must be reached at which the reaction is no longer feasible.

Reactions with a **positive entropy change**.

For example, $H_2O(g) \rightarrow H_2(g) + \frac{1}{2}O_2(g)$:

- ΔS is positive, and so $T\Delta S$ will also be positive
- As temperature increases, $T\Delta S$ becomes more positive
- $\Delta H - T\Delta S$ therefore becomes more negative as temperature increases
- ΔG therefore becomes more negative
- So the reaction remains feasible as the temperature increases.

Exam practice

1 **(a)** Define the term 'lattice formation enthalpy'. [2]

(b) Write equations that show the lattice formation enthalpy of:

(i) lithium oxide [2]

(ii) calcium chloride [2]

(c) Write these ionic compounds in order of increasing lattice dissociation enthalpy — lowest first:

NaCl, CsCl, KCl, LiCl

Explain the order that you choose. [3]

(d) Draw a Born–Haber cycle for the formation of sodium iodide. Label the species present each stage. [3]

(e) Using your Born–Haber cycle from part (d) and the data given below, calculate a value for the enthalpy of formation of sodium iodide. [2]

Data/kJ mol^{-1}:

ΔH_{latt}[NaI(s)] = −684 ΔH_{at}[Na(s)] = +109 $\Delta H_{i.e.}$[Na(g)] = +494

ΔH_{at}[iodine(g)] = +107 $\Delta H_{e.a.}$[I(g)] = −314

2 Carbon monoxide, CO, reacts with hydrogen gas, H_2, under certain conditions to form methanol, CH_3OH:

	$CO(g)$	+ $2H_2(g)$ →	$CH_3OH(l)$
S/J K^{-1} mol^{-1}	198	131	127
ΔH_f/kJ mol^{-1}	−111	0	−239

(a) Calculate the entropy change for the reaction. [2]

(b) Calculate the enthalpy change for the reaction. [2]

(c) Determine the Gibb's free energy change for the reaction at 298 K. [2]

(d) Given that the reaction is feasible at 298 K, show that the reaction ceases to be feasible at about 111°C. [2]

(e) Explain why a catalyst is added to the reaction mixture when it is used industrially. [1]

Answers and quick quizzes online

Online

Examiners' summary

You should now have an understanding of:

- ✓ the important definitions used in thermodynamics
- ✓ the Born–Haber cycle and how to construct one
- ✓ the use of lattice enthalpy (formation or dissociation) as a measure of the attractive forces between ions
- ✓ how lattice enthalpy yields information about the nature of bonding types
- ✓ how lattice enthalpy, enthalpy of solution and hydration enthalpies are related in an energy cycle
- ✓ bond enthalpies and how they are used for calculating enthalpy changes
- ✓ how bond enthalpies can either be mean values or specific values
- ✓ the meaning of entropy and how to assess the entropy change in a physical or chemical process qualitatively
- ✓ how to calculate entropy changes using absolute entropy values
- ✓ how to calculate Gibb's free energy changes using $\Delta G = \Delta H - T\Delta S$
- ✓ the significance of Gibb's free energy in terms indicating whether reactions are feasible or not
- ✓ how to determine the temperature at which a reaction may, or may not, be feasible

11 Periodicity

Periodicity is the regular and repeating pattern of physical and chemical properties when elements are arranged in order of atomic number in the periodic table.

Since the outer electronic configuration of atoms is a periodic function, we should expect other properties to change accordingly.

Periodicity can be demonstrated using the reactions and properties of the period 3 elements:

Na Mg Al Si P S Cl Ar

Reactions of period 3 elements

Reactions with water

Revised

The notable reactions of the elements with water apply largely to sodium and magnesium.

Sodium

Sodium reacts violently with water to form hydrogen gas and a solution of a **strong alkali** (sodium hydroxide) of pH 13–14. In the reaction, sodium melts, fizzes and moves around the surface of the water forming a colourless solution and colourless gas:

$$2Na(s) + 2H_2O(l) \rightarrow 2NaOH(aq) + H_2(g)$$

In the reaction, sodium atoms lose their outer electrons and are **oxidised** to form Na^+ ions:

$$Na(s) \rightarrow Na^+(aq) + e^-$$

Water molecules gain electrons, are reduced, to form hydrogen and hydroxide ions:

$$2H_2O(l) + 2e^- \rightarrow 2OH^-(aq) + H_2(g)$$

Magnesium

Magnesium reacts only very slowly with cold water to form magnesium hydroxide — a slightly soluble white solid — and a colourless gas:

$$Mg(s) + 2H_2O(l) \rightarrow Mg(OH)_2(aq) + H_2(g)$$

It reacts more rapidly with steam to form magnesium oxide, a white solid, and hydrogen gas:

$$Mg(s) + H_2O(g) \rightarrow MgO(s) + H_2(g)$$

The oxides of period 3 elements

The study of the physical and chemical properties of the oxides of the period 3 elements reveals some important trends in behaviour, which can be extended to other parts of the periodic table.

Formation of the oxides

Revised

An element can be reacted with oxygen by simply heating it in air until it ignites. The burning sample is then put into a gas jar containing pure oxygen. Alternatively, a stream of pure oxygen can be passed over a heated sample of the element in a glass tube.

A summary of the reactions of the period 3 elements with oxygen is shown in Table 11.1.

Table 11.1 Reactions of period 3 elements with oxygen

Element	Na	Mg	Al	Si	P	S
Formula and state of oxide	$Na_2O(s)$	$MgO(s)$	$Al_2O_3(s)$	$SiO_2(s)$	$P_4O_{10}(s)$	$SO_2(g)$ $SO_3(g)$
Equation	$4Na(s) + O_2(g) \rightarrow 2Na_2O(s)$	$2Mg(s) + O_2(g) \rightarrow 2MgO(s)$	$4Al(s) + 3O_2(g) \rightarrow 2Al_2O_3(s)$	$Si(s) + O_2(g) \rightarrow SiO_2(s)$	$P_4(s) + 5O_2(g) \rightarrow P_4O_{10}(s)$	$S(s) + O_2(g) \rightarrow SO_2(g)$
Observations	Melts and then ignites with an orange flame. A white solid is formed	A bright white flame is formed, and a white solid remains	Fine powder reacts to form white 'sparks'. A solid is formed as a fine white powder.	No observable reaction under normal conditions	A bright white/ yellow flame together with white fumes that collect to form a white solid	The yellow solid melts then slowly disappears as it burns with blue flame — fumes are formed
Bonding of oxide	Ionic	Ionic	Ionic with covalent character	Covalent	Covalent	Covalent
Structure of oxide	Giant ionic	Giant ionic	Giant ionic with covalent character	Giant covalent	Simple covalent	Simple covalent

Now test yourself

Tested

1 Write equations to show the oxides formed when the following elements are heated in excess oxygen. (You may want to cover up Table 11.1 to test your memory.)
- **(a)** aluminium
- **(b)** sulfur
- **(c)** sodium

2 Write an equation to show what is likely to happen when caesium, Cs, a highly reactive group 1 metal, is added to water. Comment on the pH of the solution formed.

Answers on p. 110

Reactions of period 3 oxides with water

Revised

Table 11.2 summarises the important observations made when the oxides react with water.

Table 11.2 Period 3 oxides with water

Formula of oxide	**$Na_2O(s)$**	**$MgO(s)$**	**$Al_2O_3(s)$**	**$SiO_2(s)$**	**$P_4O_{10}(s)$**	**$SO_2(g)$ $SO_3(g)$**
Reaction of oxide with water	White solid reacts rapidly forming a colourless solution, and releasing heat	A slow reaction in which some white solid dissolves	No observable change takes place to white powder	No observable change takes place to white powder	A vigorous reaction takes place; the white solid rapidly reacts to form a colourless solution	Both gases react with water to form colourless solutions
pH of solution formed	13–14	8–10	7	7	2–4	$SO_2(g)$: 3–5 $SO_3(g)$: 1–3
Formula of products	$NaOH(aq)$	$Mg(OH)_2(aq)$	—	—	$H_3PO_4(aq)$	$H_2SO_3(aq)$ $H_2SO_4(aq)$

Structure and bonding of period 3 oxides

Revised

The bonding character of the oxides changes from ionic to covalent on moving from left to right in the periodic table.

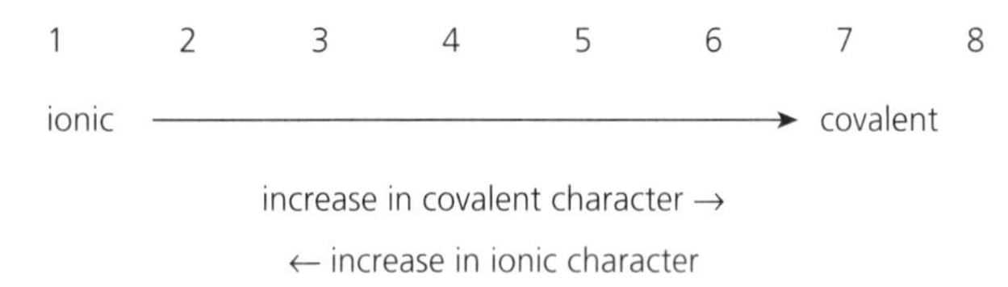

As the bonding character changes, the physical and chemical properties also change.

Physical properties of the oxides

- **Ionic oxides** like Na_2O, MgO and Al_2O_3 all have **giant ionic structures** similar to that shown in Figure 11.1. They all have high melting and boiling points because of the strong electrostatic forces acting between the oppositely charged ions. When solid, the oxides are poor electrical conductors (insulators) because the ions are not mobile, but when molten they conduct electricity well because the ions are mobile.
- **Covalent oxides** like SiO_2 have a **giant covalent structure** as shown in Figure 11.2. The silicon and oxygen atoms are bonded throughout by strong Si–O covalent bonds. These are difficult to break and hence these oxides have high melting and boiling points.

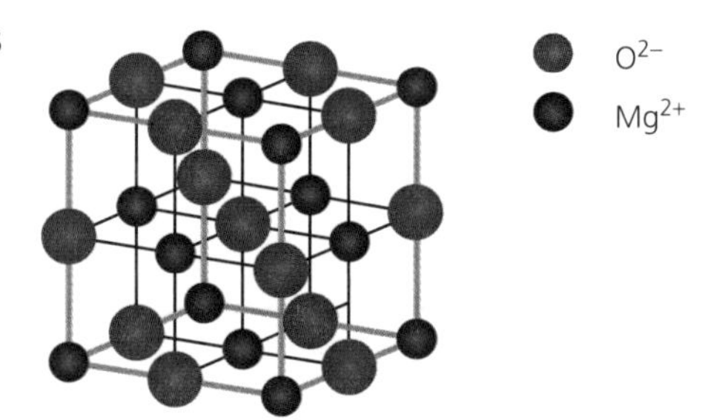

Figure 11.1 The giant ionic structure of MgO

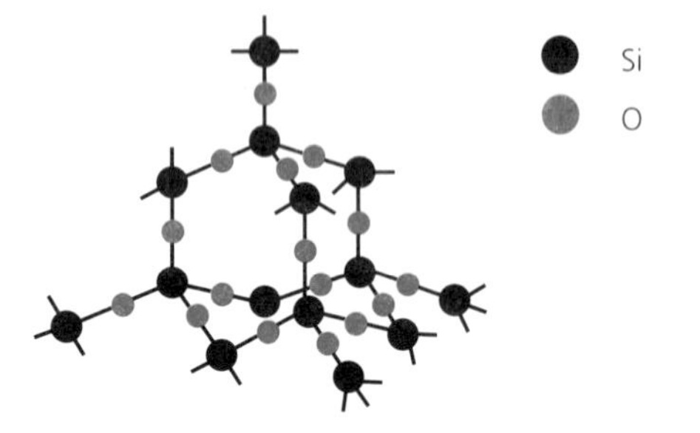

Figure 11.2 The giant structure of silicon(IV) oxide

- **Covalent oxides** like P_4O_{10}, SO_2 and SO_3 all have **simple covalent structures** and therefore exist as simple molecules with weak van der Waals' forces and dipole–dipole interactions between the molecules. Their melting and boiling points are low because these intermolecular forces are easily overcome. Electrical conductivity is poor because all molecules have a zero charge overall.

Chemical properties of period 3 oxides

The nature of the reaction of an oxide with water gives information about the bonding in the compound.

Generally, metal oxides (ionic) react with water to form hydroxide solutions with a high pH value. Non-metal oxides (covalent) form acidic solutions with low pH values. Both types of reaction involve a process called **hydrolysis**.

Hydrolysis is the splitting up of a substance by water. In this process, water molecules are chemically changed.

- Sodium oxide, $Na_2O(s)$, forms a solution of the strong alkali sodium hydroxide:

  $Na_2O(s) + H_2O(l) \rightarrow 2NaOH(aq)$

  or $O^{2-}(s) + H_2O(l) \rightarrow 2OH^-(aq)$
- Magnesium oxide, $MgO(s)$, reacts only slightly with water to form a dilute alkaline solution of magnesium hydroxide. The product is only slightly soluble in water and so forms a low concentration of hydroxide ions and is therfore a weak acid:

  $MgO(s) + H_2O(l) \rightleftharpoons Mg(OH)_2(aq)$
- Aluminium oxide is virtually insoluble in water and does not react with it.

From silicon onwards, the oxides are covalent and exist either as giant molecules (e.g. SiO_2) that are not hydrolysed by water, or as simple molecules (e.g. SO_2, P_4O_{10}) that are hydrolysed to form acidic solutions.

- Phosphorus(V) oxide, $P_4O_{10}(s)$, is hydrolysed by water to form a solution of phosphoric(V) acid:

  $P_4O_{10}(s) + 6H_2O(l) \rightarrow 4H_3PO_4(aq)$
- Sulfur(IV) oxide, $SO_2(g)$, is hydrolysed to form sulfuric(IV) acid, $H_2SO_3(aq)$ — this is a weak acid:

  $SO_2(g) + H_2O(l) \rightleftharpoons H_2SO_3(aq)$
- Sulfur(VI) oxide, $SO_3(g)$, is rapidly hydrolysed to form sulfuric(VI) acid, $H_2SO_4(aq)$ — this is a strong acid:

  $SO_3(g) + H_2O(l) \rightarrow H_2SO_4(aq)$

Acid–base properties of period 3 oxides

Because metal oxides are bases, they will react with acids to form a salt and water. Conversely, non-metal oxides are acids and will react with bases to form a salt and water.

Table 11.3 summarises the reactions. Note that aluminium oxide reacts with both acids and bases to form salts. Oxides of this type are called **amphoteric** oxides.

Now test yourself

3 Write equations to show the reactions taking place when the following oxides are added to water:
   (a) sulfur(IV) oxide
   (b) sodium oxide

4 Write an equation for the reaction between the two solutions formed in question 3.

Answers on p. 111

Tested

Table 11.3 Reactions of period 3 oxides with acids and alkalis

Oxide	Reaction with dilute HCl(aq)	Reaction with dilute NaOH(aq)
$Na_2O(s)$	$Na_2O(s) + 2HCl(aq) \rightarrow 2NaCl(aq) + H_2O(l)$	No reaction
$MgO(s)$	$MgO(s) + 2HCl(aq) \rightarrow MgCl_2(aq) + H_2O(l)$	No reaction
$Al_2O_3(s)$	$Al_2O_3(s) + 6HCl(aq) \rightarrow 2AlCl_3(aq) + 3H_2O(l)$	$Al_2O_3(s) + 2NaOH(aq) + 3H_2O(l) \rightarrow 2NaAl(OH)_4(aq)$
$SiO_2(s)$	No reaction	$SiO_2(s) + 2NaOH(aq) \rightarrow Na_2SiO_3(aq) + H_2O(l)$
$P_4O_{10}(s)$	No reaction	$P_4O_{10}(s) + 12NaOH(aq) \rightarrow 4Na_3PO_4(aq) + 6H_2O(l)$
$SO_2(g)$	No reaction	$SO_2(g) + 2NaOH(aq) \rightarrow Na_2SO_3(aq) + H_2O(l)$
$SO_3(s)$	No reaction	$SO_3(g) + 2NaOH(aq) \rightarrow Na_2SO_4(aq) + H_2O(l)$

Exam practice

1 **(a)** Describe what you would observe when magnesium burns in oxygen. Write an equation for the reaction that occurs. State the type of bonding in the oxide formed. [4]

(b) Describe what you would observe when sulfur burns in oxygen. Write an equation for the reaction that occurs. State the type of bonding in the oxide formed. [4]

(c) The substance formed in part (a) of this question is added to water forming solution A. In a separate experiment, sulfur(VI) oxide is added to water to form solution B. Solution B is then added to solution A. Write equations for these reactions:

(i) the oxide from part (a) reacting with water to form solution A [1]

(ii) sulfur(VI) oxide reacting with water to form solution B [1]

(iii) solution A reacting with solution B. [1]

(d) Outline an experiment that would show that aluminium oxide is composed of ions. [2]

(e) Write equations to show how aluminium oxide reacts with:

(i) sulfuric(VI) acid [1]

(ii) aqueous potassium hydroxide [1]

Answers and quick quizzes online

Online

Examiners' summary

You should now have an understanding of:

- ✔ the reactions of magnesium and sodium with water
- ✔ the reactions of period 3 elements with oxygen and of the oxides that are formed
- ✔ how the physical properties of the oxides depend on the bonding and structure of the compound
- ✔ the reactions of period 3 oxides with water and the meaning of the term hydrolysis
- ✔ the fact that some metal oxides form alkaline solutions, whereas some non-metal oxides form acidic solutions
- ✔ the reactions of metal oxides with acids to form salts
- ✔ the reactions of non-metal oxides with bases to form salts
- ✔ the fact that there are oxides that are amphoteric

12 Redox equilibria

Redox equations

Many reactions are classed as **redox reactions** because they involve both **reduction** and **oxidation** of different species.

Oxidation is the loss of electrons.
Reduction is the gain of electrons.

Examiners' tip

Oxidation can also be defined as the gain of oxygen and loss of hydrogen. Reduction is the loss of oxygen and gain of hydrogen.

Half-equations

Revised

In any redox reaction, one species is oxidised and another is reduced. We can write two separate half equations to show what is going on in each 'half' of the reaction in terms of electrons added or removed.

Example

When an acidified solution containing manganate(VII) ions, $MnO_4^-(aq)$, is added to aqueous iron(II) ions, a reaction occurs in which manganese(II) ions and iron(III) ions are formed. Write the two half-equations for the redox reactions taking place.

Answer

Oxidation: $Fe^{2+}(aq) \rightarrow Fe^{3+}(aq) + e^-$

For the reduction, put the equation together according to these steps:

- write down what is known: $MnO_4^- \rightarrow Mn^{2+}$
- add water, H_2O, to balance the oxygens: $MnO_4^- \rightarrow Mn^{2+} + 4H_2O$
- add hydrogen ions, H^+, to balance the hydrogens:
  $MnO_4^- + 8H^+ \rightarrow Mn^{2+} + 4H_2O$
- add electrons, e^-, to balance the charge:
  $MnO_4^- + 8H^+ + 5e^- \rightarrow Mn^{2+} + 4H_2O$

The two half-equations can then be added together to form an overall equation making sure that the electrons cancel out:

Oxidation: $Fe^{2+}(aq) \rightarrow Fe^{3+}(aq) + e^-$

Reduction: $MnO_4^- + 8H^+ + 5e^- \rightarrow Mn^{2+} + 4H_2O$

For the overall balanced equation, multiply the top half-equation by 5 and then add it to the bottom half-equation:

$5Fe^{2+}(aq) \rightarrow 5Fe^{3+}(aq) + 5e^-$

$MnO_4^- + 8H^+ + 5e^- \rightarrow Mn^{2+} + 4H_2O$

$MnO_4^-(aq) + 8H^+(aq) + 5Fe^{2+}(aq) \rightarrow Mn^{2+}(aq) + 5Fe^{3+}(aq) + 4H_2O(l)$

Examiners' tip

The numbers of electrons in each half-equation must be the same before they are added together.

Oxidation states

Revised

An **oxidation state** of an element can be thought of as the 'formal' charge that the element has when combined in a compound. For example, in $CrCl_3$ the oxidation state of the chlorine is –1 and that of the chromium is +3 (the sum of the individual oxidation states is zero for a compound).

Transition metal ions can often exist in more than one oxidation state. We must indicate in the name of a compound just what the oxidation state of the metal is. For example, we use names such as copper(II) sulfate (with Cu^{2+} ions), copper(I) oxide (with Cu^{+} ions) and chromium(III) fluoride (with Cr^{3+} ions).

If an **oxidation state** increases in a reaction, then oxidation has occurred (loss of electrons) and reduction has occurred if an oxidation state has decreased.

Now test yourself

Tested

1 When dichromate(VI) ions, $Cr_2O_7^{2-}$(aq), are added to an acidified solution containing iodide ions, I^-(aq), a redox reaction occurs in which an aqueous solution containing chromium(III) ions and iodine is formed.
  - **(a)** Write the two half-equations for the reaction.
  - **(b)** Which substances have been (i) reduced and (ii) oxidised in the reaction?
  - **(c)** Write the overall equation for the reaction by combining the two half-equations from part (a).
  - **(d)** What are the oxidation state changes of (i) chromium and (ii) iodine in the reaction?

2 A solution containing the ferrate(VI) ion, FeO_4^{2-}(aq) is added to an acidified solution containing manganese(IV) oxide powder. An aqueous solution containing iron(III) ions and manganate(VII) ions is formed.
  - **(a)** Write the two half-equations for the reaction.
  - **(b)** Which substance has been (i) reduced and (ii) oxidised in the reaction?
  - **(c)** Write the overall equation by combining the two half-equations from part (a).
  - **(d)** What are the oxidation state changes of (i) manganese and (ii) iron in the reaction?

Answers on p. 111

Electrode potentials

Measuring a standard electrode potential

Revised

A **standard reference electrode** — a hydrogen electrode — is used to measure a **standard electrode potential**. The potential difference of this electrode under standard conditions is defined as 0.00 volts.

Figure 12.1 shows how the standard electrode potential is measured for the Cu^{2+}(aq) + $2e^- \rightleftharpoons$ Cu(s) system. Note that half-equations for electrode systems are always written as reductions.

The **standard electrode potential** of a substance is the potential difference, in volts, of a substance in an aqueous solution of its ions relative to the standard hydrogen electrode, under standard conditions: 298 K, 100 kPa and 1 mol dm^{-3} solutions.

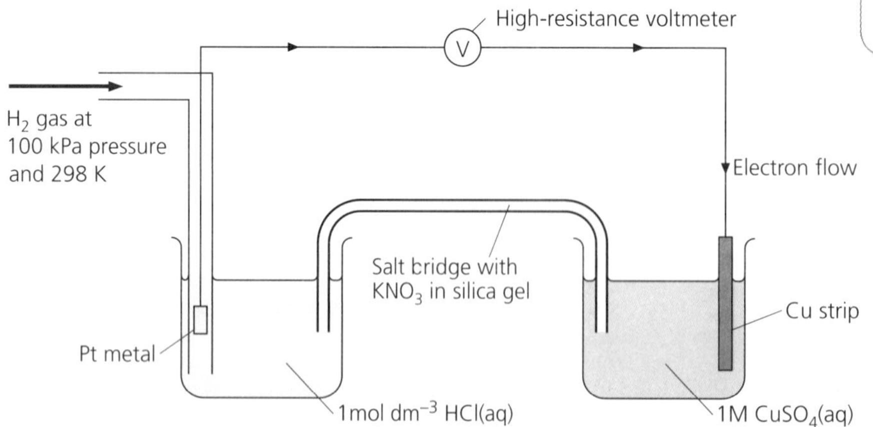

Figure 12.1 Measuring an electrode potential

Two half-cells are shown — the one on the left is a standard hydrogen electrode and the one on the right shows a piece of copper in contact with Cu^{2+} ions. A high-resistance voltmeter is used so that it reads the potential difference with, effectively, no current passing through it.

Typical mistake

If asked to draw a diagram showing how the standard electrode potential for $Fe^{3+}(aq) + e^- \rightleftharpoons Fe^{2+}(aq)$ is measured, it is important to include 1.00 mol dm^{-3} solutions of *both* Fe^{2+} ions *and* Fe^{3+} ions in contact with a platinum (inert) electrode.

A reading of +0.34 volts is recorded from the cell terminals as measured in Figure 12.1. So we say that the electrode potential for the $Cu^{2+}(aq) + 2e^- \rightleftharpoons Cu(s)$ system is +0.34 V under standard conditions.

The two half-equations (written as reduction processes) are:

Hydrogen electrode: $2H^+(aq) + 2e^- \rightleftharpoons H_2(g)$; $E^\ominus = 0.00\ V$

Copper electrode: $Cu^{2+}(aq) + 2e^- \rightleftharpoons Cu(s)$; $E^\ominus = +\ 0.34\ V$

Some points to note:

- $E^\ominus$ is used to indicate an electrode potential measured under standard conditions.
- The more positive electrode potential shifts to the right-hand side:
  $Cu^{2+}(aq) + 2e^- \rightarrow Cu(s)$
- The more negative electrode potential shifts to the left-hand side
  $H_2(g) \rightarrow 2H^+(aq) + 2e^-$
- Electrons are gained by the copper(II) ions at the copper electrode, which will therefore have a positive charge. Reduction occurs at this electrode.
- Electrons are released at the hydrogen electrode, which will therefore have a negative charge. Oxidation takes place at this electrode.
- Electrons move from the hydrogen half-cell to the copper half-cell in the external circuit — from the negative terminal to the positive terminal.
- Adding the two half-equations together gives the overall equation:
  $Cu^{2+}(aq) + H_2(g) \rightarrow Cu(s) + 2H^+(aq)$
- The overall electrode potential is equal to the two individual processes added together, but the reverse process has its sign changed:
  $E^\ominus = +0.34\ V + 0.00\ V = +\ 0.34\ V$.
- As the overall electrode potential is positive, this means that the reaction as written is spontaneous, that is
  $Cu^{2+}(aq) + H_2(g) \rightarrow Cu(s) + 2H^+(aq)$

Example

Some reduction processes, together with their electrode potentials, are written below:

(1) $Zn^{2+}(aq) + 2e^- \rightleftharpoons Zn(s)$; $E^\ominus = -0.76\ V$

(2) $I_2(aq) + 2e^- \rightleftharpoons 2I^-(aq)$; $E^\ominus = +0.54\ V$

(3) $Fe^{3+}(aq) + e^- \rightleftharpoons Fe^{2+}(aq)$; $E^\ominus = +0.77\ V$

Combine the following pairs of equations, giving the overall electrode potentials for the spontaneous processes:

(a) (1) and (2)

(b) (2) and (3)

Answer

(a) Equilibrium (2) shifts to the right-hand side because it is more positive; equilibrium (1) shifts to the left-hand side:

(1) $Zn^{2+}(aq) + 2e^- \rightleftharpoons Zn(s)$; $E^\ominus = -0.76$ V ←

(2) $I_2(aq) + 2e^- \rightleftharpoons 2I^-(aq)$; $E^\ominus = +0.54$ V →

Reversing equilibrium 1 and adding gives:

$$Zn(s) + I_2(aq) \longrightarrow Zn^{2+}(aq) + 2I^-(aq)$$

The overall electrode potential for this spontaneous reaction will be:

$E^\ominus = +0.54 + (+0.76)$

$= +1.30$ V

The sign of the electrode potential is positive so the reaction in which zinc reduces iodine to iodide ions, and zinc atoms are oxidised to zinc ions, is spontaneous.

(b) Equilibrium 3 shifts to the right-hand side — it is more positive; equilibrium 2 shifts to the left-hand side:

(2) $I_2(aq) + 2e^- \rightleftharpoons 2I^-(aq)$; $E^\ominus = +0.54$ V ←

(3) $Fe^{3+}(aq) + e^- \rightleftharpoons Fe^{2+}(aq)$; $E^\ominus = +0.77$ V →

Reversing equilibrium 2 and multiplying equilibrium 3 by 2 to balance the electrons and then adding gives:

$$2Fe^{3+}(aq) + 2I^-(aq) \longrightarrow 2Fe^{2+}(aq) + I_2(aq)$$

The overall electrode potential for this reaction, the spontaneous one, will be:

$E^\ominus = +0.77 + (-0.54)$

$= +0.23$ V

So, iron(III) ions will oxidise iodide ions to form iodine and iron(II) ions.

Now test yourself Tested

3 **Predict the products of the reactions, if any, that occurs when aqueous iron(III) ions are added to solutions containing chloride ions, bromide ions and iodide ions in three separate experiments. Use the electrode potentials:**

$Cl_2(aq) + 2e^- \rightleftharpoons 2Cl^-(aq)$; $E^\ominus = +1.36$ V

$Br_2(aq) + 2e^- \rightleftharpoons 2Br^-(aq)$; $E^\ominus = +1.07$ V

$I_2(aq) + 2e^- \rightleftharpoons 2I^-(aq)$; $E^\ominus = +0.54$ V

$Fe^{3+}(aq) + e^- \rightleftharpoons Fe^{2+}(aq)$; $E^\ominus = +0.77$ V

Answers on p. 111

Electrochemical series

Standard electrode potentials can be written as a list called the **electrochemical series** in which each electrochemical process is written as a reduction process.

Table 12.1 gives an example of an electrochemical series.

Table 12.1 An electrochemical series

Oxidising power of left-hand species increases ↑		$E^{\ominus}$V	
	$F_2(g) + 2e^- \rightarrow 2F^-(aq)$	+2.87	
	$H_2O_2(aq) + 2H^+(aq) + 2e^- \rightarrow 2H_2O(1)$	+1.77	
	$Au^+(aq) + e^- \rightarrow Au(s)$	+1.68	
	$Cl_2(g) + 2e^- \rightarrow 2Cl^-(aq)$	+1.36	
	$O_2(g) + 4H^+(aq) + 4e^- \rightarrow 2H_2O(1)$	+1.23	
	$Br_2(l) + 2e^- \rightarrow 2Br^-(aq)$	+1.09	
	$Ag^+(aq) + e^- \rightarrow Ag(s)$	+0.80	
	$Fe^{3+}(aq) + e^- \rightarrow Fe^{2+}(aq)$	+0.77	
	$I_2(s) + 2e^- \rightarrow 2I^-(aq)$	+0.54	
	$O_2(g) + 2H_2O(1) + 4e^- \rightarrow 4OH^-(aq)$	+0.40	
	$Cu^{2+}(aq) + 2e^- \rightarrow Cu(s)$	+0.34	
	$S(s) + 2H^+(aq) + 2e^- \rightarrow H_2S(g)$	+0.14	
	$2H^+(aq) + 2e^- \rightarrow H_2(g)$	0.00	
	$Pb^{2+}(aq) + 2e^- \rightarrow Pb(s)$	−0.13	
	$Sn^{2+}(aq) + 2e^- \rightarrow Sn(s)$	−0.14	
	$Ni^{2+}(aq) + 2e^- \rightarrow Ni(s)$	−0.25	
	$Co^{2+}(aq) + 2e^- \rightarrow Co(s)$	−0.28	↓ Reducing power of right-hand species increases

Notice the following about this series:

- Species on the left are oxidising agents.
- Species on the right are reducing agents.
- The electrode potentials increase in positive value going up the list.
- This means that the oxidising power of the species on the left of the equilibrium at the top of the list (F_2) is greater than all those below it.
- An oxidising agent at the top of the list (on the left) will oxidise any reducing agent below it (on the right). For example fluorine gas, $F_2(g)$, will oxidise anything below it on the right-hand side — for example $Br^-(aq)$, $Ag(s)$, $Fe^{2+}(aq)$, $I^-(aq)$, $OH^-(aq)$, $Cu(s)$ and $H_2S(s)$.
- The reducing power of the species on the right increases going down the list.
- A species like Co(s), Ni(s), Sn(s) and Pb(s) will all reduce any species on the left-hand side above it in the list.

Representation of electrochemical cells

Revised

For all electrochemical cells (combinations of two half-cells), a **conventional representation** can be written that includes the redox processes taking place in the cell. This cell 'diagram' is a symbolic form representing the electrochemical features and processes taking place. Figure 12.2 shows this conventional representation.

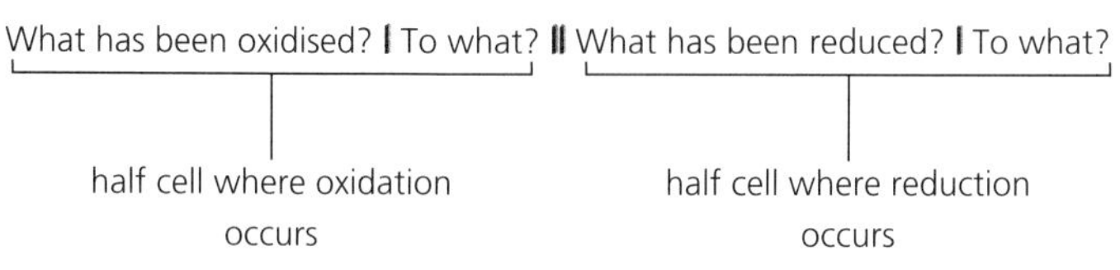

Figure 12.2 Representing an electrochemical cell

The single vertical straight lines represent phase-change boundaries — for example, a solid in contact with a solution. The double vertical line represents the salt bridge through which ions flow.

Calculating overall cell potentials

A cell potential is also called an electromotive force (e.m.f.). An overall cell potential, $E^\ominus_{cell}$, is calculated using:

$$E^\ominus_{cell} = E^\ominus_{rhs} - E^\ominus_{lhs}$$

where $E^\ominus_{rhs}$ and $E^\ominus_{lhs}$ are the two electrode potentials for the half-equations written as reductions.

Examiners' tip

If an overall electrode potential is positive then the redox process being considered is feasible, but the rate is not predictable.

Example

Consider these two half-equations written as reduction potentials:

$Ag^+(aq) + e^- \rightleftharpoons Ag(s)$; $E^\ominus = +0.80\,V$

$Ni^{2+}(aq) + 2e^- \rightleftharpoons Ni(s)$; $E^\ominus = -0.25\,V$

- Because the Ag^+/Ag electrode potential is more positive than the Ni^{2+}/Ni electrode potential, the silver process will proceed from left to right as written above. Electrons are required in this process and these are gained from the nickel metal. This forces the nickel half-reaction to move from right to left. So the spontaneous processes will be:

  $Ag^+(aq) + e^- \rightarrow Ag(s)$

  and this is the positive electrode because electrons are being removed from it;

  $Ni(s) \rightarrow Ni^{2+}(aq) + 2e^-$

  and this is the negative electrode because electrons are being released.
- Silver ions are therefore reduced to silver, and nickel is oxidised to nickel(II) ions.
- Figure 12.3 shows the conventional representation for the spontaneous process with the polarity of the terminals shown.

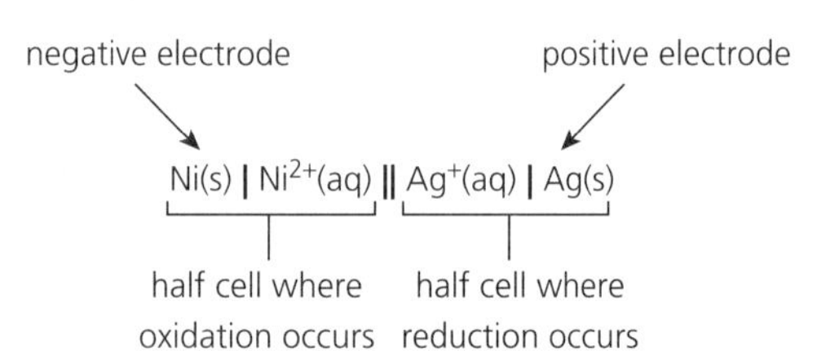

Figure 12.3 Cell representation

- The overall electrode potential, $E^\ominus_{cell}$, is calculated as follows:

  $E^\ominus_{cell} = E^\ominus_{rhs} - E^\ominus_{lhs}$

  $= +0.80 - (-0.25)$

  $= +1.05\,V$

Examiners' tip

In a cell diagrams, each half must balance but — it is not necessary to balance the charge by writing a '2' in front of the silver ion and the silver atom on the right in Figure 12.3.

Now test yourself Tested

4 Given the electrochemical series below:

	$E^{\ominus}$/V
$F_2(g) + 2e^- \rightarrow 2F^-(aq)$	+2.87
$H_2O_2(aq) + 2H^+(aq) + 2e^- \rightarrow 2H_2O(l)$	+1.77
$Au^+(aq) + e^- \rightarrow Au(s)$	+1.68
$Cl_2(g) + 2e^- \rightarrow 2Cl^-(aq)$	+1.36
$O_2(g) + 4H^+(aq) + 4e^- \rightarrow 2H_2O(l)$	+1.23
$Br_2(l) + 2e^- \rightarrow 2Br^-(aq)$	+1.09

Indicate whether each of the following pairs could result in a reaction or not.

(a) **(i)** $Br_2(l)$ and $Cl^-(aq)$
(ii) $H_2O_2(aq)$ in acidic solution and $Br^-(aq)$
(iii) $Au^+(aq)$ and $Br_2(l)$
(iv) $F^-(aq)$ and $Br_2(l)$
(v) $O_2(g)$ in acidic solution and $Br^-(aq)$

(b) A cell is set up involving the electrode systems Au^+/Au ($E^{\ominus}$ = +1.68 V) and Zn^{2+}/Zn ($E^{\ominus}$ = −0.76 V).
(i) Draw a labelled diagram of the cell formed when each half-cell is combined.
(ii) Calculate the overall cell potential.
(iii) Give the conventional representation of this cell.

Answers on p. 111

Electrochemical cells

In an electrochemical cell, there takes place an electrochemical process from which electrical energy can be obtained, and then used for doing useful work. All electrochemical cells contain two halves — oxidation takes place in one half and reduction in the other.

Many cells are **non-rechargeable** because their electrochemical process is irreversible. However, other cells use processes that are reversible and these can be recharged.

A lead storage cell is an example of a reversible process, so these cells can be recharged. The half-reactions taking place at the electrodes during discharge (the spontaneous process) are as follows:

$$PbO_2(s) + 3H^+(aq) + HSO_4^-(aq) + 2e^- \rightleftharpoons PbSO_4(s) + 2H_2O(l)$$
$$Pb(s) + HSO_4^-(aq) \rightleftharpoons PbSO_4(s) + H^+(aq) + 2e^-$$

Adding these two together gives the overall process taking place:

$$Pb(s) + PbO_2(s) + 2H_2SO_4(aq) \rightleftharpoons 2PbSO_4(s) + 2H_2O(l)$$

The forward reaction describes the spontaneous discharging process and the reverse reaction describes the charging process.

Fuel cells

Revised

In a typical electrochemical cell, the potential difference decreases with time because the concentrations in the half-cells change as the spontaneous cell reaction takes place. This continues until the electrochemical processes are in equilibrium, and the cell can no longer be used.

In a **fuel cell**, the 'fuel' can be added constantly so that the current is constant with time. The cell therefore does not need charging all the time.

Hydrogen–oxygen fuel cells have been shown to be very useful. In this cell, hydrogen is the fuel and it undergoes the process of oxidation at the anode. Electrons are released that then pass round the external circuit, do work, and are then used to reduce oxygen molecules. These reactions can happen in either acidic or alkaline conditions:

- in acid:

  anode: $2H_2(g) \rightarrow 4H^+(aq) + 4e^-$

  cathode: $O_2(g) + 4H^+(aq) + 4e^- \rightarrow 2H_2O(l)$
- in alkali:

  anode: $2H_2(g) + 4OH^-(aq) \rightarrow 4H_2O(l) + 4e^-$

  cathode: $O_2(g) + 2H_2O(l) + 4e^- \rightarrow 4OH^-(aq)$

As usual, the overall process can be obtained by adding the half-equations for the individual electrochemical processes so that the electrons cancel:

$$2H_2(g) + O_2(g) \rightarrow 2H_2O(l)$$

Examiners' tip

In examinations, you may be asked to add together two half-equations to form an overall equation. Alternatively you may be asked to work out a missing half-equation if you are provided with the overall equation and the other half-equation.

Advantages and disadvantages of fuel cells

Advantages

Fuel cells:

- are much more energy-efficient than conventional methods of energy provision
- are less complex than conventional gas or diesel engines
- are not subject to high temperatures or structural weaknesses found in other engines
- will operate indefinitely as long as the fuel is available
- produce no pollutants — the only emission from a hydrogen–oxygen fuel cell is water.

Disadvantages

- Hydrogen requires costly and fairly inefficient methods of extraction from existing sources. Its production could also produce carbon dioxide as one of the by-products.
- Despite the fact that hydrogen is a low-density gas and any leaks can be dissipated quickly, there remains a serious risk of fire and, potentially, explosion.
- Storing hydrogen is a problem — the fuel tank would have to be large, under pressure and thermally insulated (hydrogen boils at –252°C). The gas could possibly be absorbed into a material like palladium, or adsorbed onto some metallic surfaces.

Exam practice

1 Redox reactions occur during the discharge of all electrochemical cells. Some of these cells are of commercial value.

The table below gives some redox half-equations and standard electrode potentials.

Half-equation	$E^{\ominus}$/V
$Cr^{2+}(aq) + 2e^- \rightleftharpoons Cr(s)$	−0.91
$AgCl(s) + e^- \rightleftharpoons Ag(s) + Cl^-(aq)$	+0.22
$O_2(g) + 4H^+(aq) + 4e^- \rightleftharpoons 2H_2O(l)$	+1.23
$O_3(g) + 2H^+(aq) + 2e^- \rightleftharpoons O_2(g) + H_2O(l)$	+2.07

(a) In terms of electrons, state what is meant by a 'reducing agent'. [1]

(b) Use the table above to identify the strongest reducing agent from the species listed. Explain your answer. [2]

(c) Use data from the table to explain why ozone in acidic solution reacts with chromium. Write an equation for the reaction that occurs. [3]

(d) An electrochemical cell can be constructed using a chromium electrode and an electrode in which silver and chloride ions are in contact with silver chloride. The cell can be used to power certain electronic devices.

(i) Give the conventional representation for this cell. [1]

(ii) Calculate the e.m.f. of the cell. [1]

(iii) Write the overall equation for the reaction that occurs during discharge. [1]

(iv) Suggest one reason why the cell cannot be electrically recharged. [1]

Answers and quick quizzes online

Online

Examiners' summary

You should now have an understanding of:

- what is meant by a 'redox' reaction
- how to use oxidation states
- how to write half-equations for redox processes and how to combine them to form an overall equation
- conventional representations of electrochemical cells
- the standard hydrogen electrode and how it is used to measure standard electrode potentials
- what standard conditions are and how important they are when measuring standard electrode potentials
- the electrochemical series and how it can be used to predict the direction of spontaneous chemical change
- some important electrochemical cells — rechargeable, non-rechargeable and fuel cells
- the hydrogen–oxygen fuel cell and the electrochemical processes involved
- the lead–acid storage cell as an example of a rechargeable cell
- the advantages and disadvantages of fuel cells

13 Transition metals

Electronic configurations

The first series of **transition metals** is:

$_{21}Sc$ $_{22}Ti$ $_{23}V$ $_{24}Cr$ $_{25}Mn$ $_{26}Fe$ $_{27}Co$ $_{28}Ni$ $_{29}Cu$ $_{30}Zn$

Although all transition metals are in the d-block of the periodic table, not all d-block elements are transition metals — for example zinc is not.

A **transition metal** is defined as a metal which, in at least one of its stable ions, has a partly-filled d-sub-level.

Electronic configurations of atoms

Revised

When writing the electronic configurations of these elements, the order in which the sub-levels fill up is important:

1s, 2s, 2p, 3s, 3p, 4s, 3d

Notice that the 4s sub-level fills before the 3d sub-level.

Examiners' tip

Zinc is not regarded as a transition metal because it forms only one stable ion, Zn^{2+}, in which it has a completely filled 3d sub-level.

Example

Write the electronic configurations of titanium, manganese and nickel.

Answer

Filling up the sub-levels in order of energy, lowest first, gives:

$_{22}Ti$ — $\mathbf{1s^2, 2s^2, 2p^6, 3s^2, 3p^6, 4s^2, 3d^2}$

$_{25}Mn$ — $\mathbf{1s^2, 2s^2, 2p^6, 3s^2, 3p^6, 4s^2, 3d^5}$

$_{28}Ni$ — $\mathbf{1s^2, 2s^2, 2p^6, 3s^2, 3p^6, 4s^2, 3d^8}$

However, note that there are two exceptions, copper and chromium, to the pattern that two electrons always go into the 4s sub-level first.

$_{29}Cu$ — $1s^2, 2s^2, 2p^6, 3s^2, 3p^6, 4s^1, 3d^{10}$

$_{24}Cr$ — $1s^2, 2s^2, 2p^6, 3s^2, 3p^6, 4s^1, 3d^5$

This happens because of the stability of full and, to a lesser extent, half-filled d-sub-levels.

Examiners' tip

Remember that chromium atoms and copper atoms have only one electron in their 4s sub-level or $4s^1$, $3d^x$. The other transition metals have two — that is $4s^2$, $3d^x$.

Electronic configurations of ions

Revised

General properties of transition metals

A maximum of 10 electrons can be added to a d-sub-level but, as can be seen from the electronic configurations described previously, many atoms and ions have a partially filled d-sub-level — that is, fewer than 10 electrons.

Many of the special **properties** of transition metals arise from an incomplete d-sub-level in atoms or ions.

Typical mistake

When writing the electronic configuration of *ions*, it must be remembered that the electrons are removed from the 4s sub-level first. [4s are first in and first out]. This is a common mistake made in examinations.

Example

Write the electronic configuration of the ions:

Fe^{3+}; Mn^{2+}; Cr^{3+}; Cu^{2+}

Answer

Step 1: write the electronic configuration of the atoms:

Fe $1s^2, 2s^2, 2p^6, 3s^2, 3p^6, 4s^2, 3d^6$

Mn $1s^2, 2s^2, 2p^6, 3s^2, 3p^6, 4s^2, 3d^5$

Cr $1s^2, 2s^2, 2p^6, 3s^2, 3p^6, 4s^1, 3d^5$

Cu $1s^2, 2s^2, 2p^6, 3s^2, 3p^6, 4s^1, 3d^{10}$

Step 2: Remove the electrons required to form the ion charge (4s out first):

Fe^{3+} $1s^2, 2s^2, 2p^6, 3s^2, 3p^6, 3d^5$

Mn^{2+} $1s^2, 2s^2, 2p^6, 3s^2, 3p^6, 3d^5$

Cr^{3+} $1s^2, 2s^2, 2p^6, 3s^2, 3p^6, 3d^3$

Cu^{2+} $1s^2, 2s^2, 2p^6, 3s^2, 3p^6, 3d^9$

Many of the properties of properties of transition metal atoms and ions are explained by their partially filled d-sub-levels:

- formation of complexes
- formation of coloured ions
- variable oxidation states
- catalytic behaviour.

Complexes

Revised

Transition metal ions can combine with molecules or ions to form **complexes**. A complex involves one or more ligands that have formed **dative** or **co-ordinate** bonds to a central metal cation.

A **ligand** is a molecule or ion with the capacity to donate an electron pair and form a co-ordinate (dative) bond.

A **ligand** is a molecule or ion with the capacity to donate an electron pair to a metal ion and form a dative bond.

Some ligands form just one co-ordinate bond per ligand — these are called **unidentate** or **monodentate** ligands. Water, ammonia, chloride ions, cyanide ions are all monodentate ligands.

Other ligands can form more than one bond per ligand:

- Those that form two co-ordinate bonds per ligand are called **bidentate** ligands — examples (Figure 13.1) include ethane-1,2-diamine ('en'), $H_2N(CH_2)_2NH_2$ and the ethanedioate ion, $C_2O_4^{2-}$.

ethane-1,2-diamine (en)

ethanedioate ion

Figure 13.1 Bidentate ligands

- Those that form more than two co-ordinate bonds per ligand are called **multidentate** ligands. Figure 13.2 shows a derivative of EDTA, which can form six co-ordinate bonds. This ligand is normally used at high pH so that all four carboxylic acid groups are deprotonated (as shown) and the ligand is $[EDTA]^{4-}$.

Figure 13.2 A multidentate ligand

Co-ordination number

The number of co-ordinate bonds formed between a metal ion and its ligands is called the ion's **co-ordination number**.

- In the complex shown in Figure 13.3, in which four chloride ions, Cl^-, act as ligands to a central copper atom, four co-ordinate bonds are formed, so the co-ordination number is 4.

or $[CuCl_4]^{2-}$

Figure 13.3 Co-ordination number 4

- Small ligands like water and ammonia, normally form octahedral complexes in which the co-ordination number of the complex is 6. In the complex in Figure 13.4, six cyanide ions, CN^-, act as ligands and form an octahedral complex with an iron(III) ion.

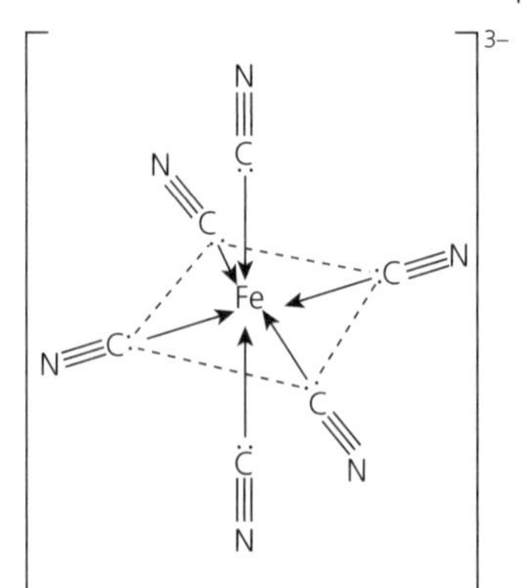

Figure 13.4 Co-ordination number 6

- In the complex shown in Figure 13.5, there is only one ligand, $EDTA^{4-}$, but it has formed six co-ordinate bonds to the central metal ion, M^+. So, the co-ordination number is 6.

Figure 13.5 Co-ordination number 6

- If a ligand is large then there may be only sufficient room for a maximum number of ligands. For example, the complex drawn in Figure 13.6 has a co-ordination number of 4 and is tetrahedral.

Typical mistake

The co-ordination number is *not* the same as the number of ligands. This is only true if the ligands are unidentate. The co-ordination number is the number of co-ordinate bonds formed by the transition metal ion.

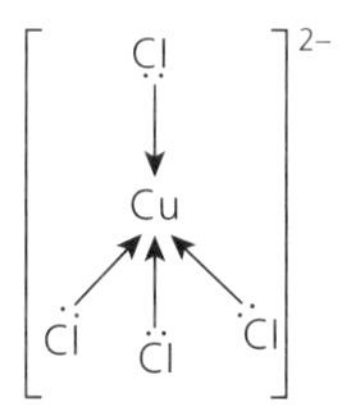

Figure 13.6 Co-ordination number 4

- Other complexes can form with a co-ordination number of 4, but with the ligands arranged in a **square planar** shape, as opposed to tetrahedral. Figure 13.7 shows a complex called cis-platin (an anti-cancer compound). The complex binds with some of the nitrogen atoms in the bases within DNA and prevents the further replication that results in cancerous growths. However, the compound is highly toxic and will also affect other parts of the body if concentrations and targets are not controlled carefully.

H_3N Cl
Pt
H_3N Cl

Figure 13.7 Cis-platin

- Figure 13.8 shows that the complex of a silver ion and ammonia molecules is linear with a co-ordination number of 2. It is used in Tollens' reagent and forms a silver mirror when silver ions are reduced to silver atoms when warmed with an aldehyde (but not with a ketone).

$$[H_3N{:} \longrightarrow Ag \longleftarrow {:}NH_3]^+$$

Figure 13.8 The active ingredient in Tollens' reagent

- Figure 13.9 shows haem, the compound responsible for forming the red pigment in blood. It consists of a porphyrin ring structure forming four co-ordinate bonds with a central iron atom. The co-ordination number is 4, so there are two lone pairs available for carrying oxygen.

CH_2 CH_3
H_3C H C CH_2
N N
HC Fe CH
N N
H_3C CH_3
C H
HOOC COOH

Figure 13.9 Haem — the oxygen carrier

This complex enables oxygen to be transported around the body by having available sites on the iron atom for oxygen molecules to act as ligands and be 'carried off' in the bloodstream. However, if carbon monoxide is inhaled, this molecule also acts as a ligand and binds strongly to the iron atom forming carboxyhaemoglobin. In doing so, the available sites for oxygen transportation are reduced and this then results in less oxygen reaching important organs in the body. For this reason, carbon monoxide is toxic.

Examiners' tip

When ligands are relatively large, fewer can 'fit' around the central metal ion.

Now test yourself Tested

1 What is the co-ordination number of the metal ion in each of these complexes?
   (a) $[CuCl_4]^{2-}$
   (b) $[Fe(H_2O)_6]^{3+}$
   (c) $[Pt(NH_3)_2Cl_2]$
   (d) $[Ni(EDTA)]^{2-}$
   (e) $[Cr(en)_3]^{3+}$

2 A complex has the formula $[Cr(C_2O_4)_3]^{3-}$.
   (a) What is the name of the ligand in the complex?
   (b) What is the shape of the complex?
   (c) What is the oxidation state of the chromium?
   (d) The same ligand reacts with nickel(II) ions. Suggest the formula of the complex formed.

3 The ligand below is called indolo[2,3-*b*]carbazole, abbreviated to 'ind'.

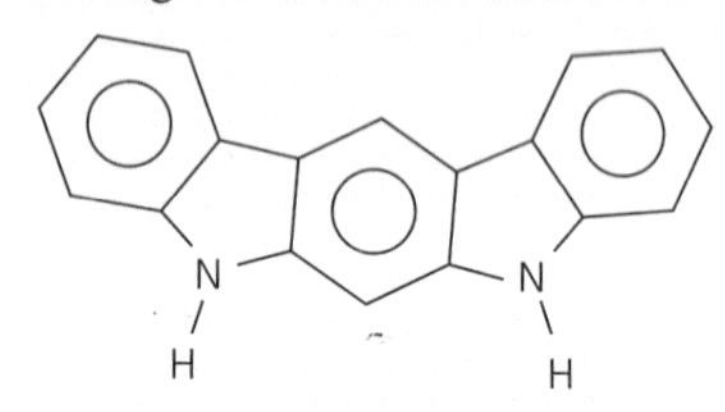

indolo[2,3-b]carbazole

   (a) How many bonds would one 'ind' molecule be expected to form with a transition metal ion?
   (b) Using the abbreviation 'ind', write the formula of the octahedral complex formed when 'ind' bonds with iron(II) ions.

Answers on p. 111

Examiners' tip

Try to learn the different shapes in terms of co-ordination numbers: octahedral = 6; tetrahedral and square planar = 4; linear = 2 etc.

Formation of coloured ions

Revised

Many transition metal compounds are coloured, especially in aqueous solution, because they contain coloured complexes:

- $[Co(H_2O)_6]^{2+}$ is pink
- $[Cr(H_2O)_6]^{3+}$ is green
- $[Fe(H_2O)_6]^{2+}$ is pale green
- $[Fe(H_2O)_6]^{3+}$ is yellow/brown
- $[Cu(H_2O)_6]^{2+}$ is blue
- $[CuCl_4]^{2-}$ is yellow
- $[CoCl_4]^{2-}$ is blue
- $[Ag(NH_3)_2]^{+}$ is colourless.

There are several factors to consider.

- Colours may change if any or all of oxidation state, co-ordination number or ligand change.
- Colour arises when visible light is absorbed by a complex and electrons are promoted from the ground state to a higher energy state.
- The energy difference between the ground state and higher state, ΔE, involves a photon of frequency ν such that:

  $$\Delta E = h\nu$$

  where h is the Planck constant.
- The light that is transmitted consists of the frequencies that are not absorbed by the complex.

- For metal ions that have either an empty or completely full d-sub-level, no electronic transitions are possible and therefore no absorption of visible light will occur. The resulting complex will be colourless.

The concentration of transition metal ions in solution can be determined by their absorbance of visible light. The higher the concentration of the ions, the greater the absorbance. Visible spectrometry can be used to make such a determination.

Variable oxidation states

Revised

Many transition elements show variable oxidation states in their compounds. For example, iron can be +2 or +3; copper can be +1 or +2; manganese can be +2, +4, +6 or +7 in their common oxidation states.

Chromium

Chromium can exist in the +2, +3 and +6 oxidation states. The table shows the colours these oxidation states normally have in aqueous solution.

Oxidation state	+2	+3	+6	
Aqueous complex	$[Cr(H_2O)_6]^{2+}(aq)$	$[Cr(H_2O)_6]^{3+}(aq)$	$CrO_4^{2-}(aq)$ or	$Cr_2O_7^{2-}(aq)$
Colour	Blue	Green	Yellow	Orange

Chromium(III) and chromium(II) compounds can be made by reducing dichromate(VI) ions in acidic conditions with zinc acting as the reducing agent:

- zinc is oxidised: $Zn \rightarrow Zn^{2+} + 2e^-$
- chromium in dichromate(VI) is reduced:
  $Cr_2O_7^{2-} + 14H^+ + 6e^- \rightarrow 2Cr^{3+} + 7H_2O$

Multiplying the zinc oxidation equation by 3, and adding gives:

$$Cr_2O_7^{2-}(aq) + 14H^+(aq) + 3Zn(s) \rightarrow 2Cr^{3+}(aq) + 3Zn^{2+}(aq) + 7H_2O(l)$$

Chromium(II) is then formed by further reduction with zinc:

- zinc is oxidised: $Zn \rightarrow Zn^{2+} + 2e^-$
- chromium(III) is reduced: $Cr^{3+} \rightarrow Cr^{2+} + e^-$

Multiplying the chromium(III) equation by 2, and adding gives:

$$2Cr^{3+}(aq) + Zn(s) \rightarrow 2Cr^{2+}(aq) + Zn^{2+}(aq)$$

The **chromate(VI)** ion, CrO_4^{2-}, can be made by oxidising chromium(III) with hydrogen peroxide in alkaline conditions. During the reaction, the colour changes from green to bright yellow:

$$2[Cr(OH)_6]^{3-}(aq) + 3H_2O_2(aq) \rightarrow 2CrO_4^{2-}(aq) + 8H_2O(aq) + 2OH^-(aq)$$

The yellow chromate(VI) ion, CrO_4^{2-}, can be converted into the orange dichromate(VI) ion, $Cr_2O_7^{2-}$, by adding acid:

$$2\,CrO_4^{2-}(aq) + 2H^+(aq) \rightleftharpoons Cr_2O_7^{2-}(aq) + H_2O(l)$$

- Adding acid, H^+, shifts the equilibrium to the right-hand side and forms the orange dichromate(VI) ion, $Cr_2O_7^{2-}$
- Adding alkali, OH^-, shifts the equilibrium to the left-hand side and yellow chromate(VI) reforms.

Examiners' tip

The dichromate–chromate interconversion is not a redox reaction because the oxidation state of the chromium before and after the reaction is the same, +6.

Iron

Iron can either be in the +2 or +3 oxidation states. They exist in solution in the form of octahedral aqua-complex ions.

Oxidation state	+2	+3
Aqueous complex	$[Fe(H_2O)_6]^{2+}$	$[Fe(H_2O)_6]^{3+}$
Colour	Pale green	Yellow/brown

Iron(II) ions are readily converted into **iron(III)** ions using oxidising agents such as acidified dichromate(VI) or manganate(VII) ions:

- iron(II) is oxidised to iron(III): $Fe^{2+} \rightarrow Fe^{3+} + e^-$
- chromium in dichromate(VI) is reduced:
  $Cr_2O_7^{2-} + 14H^+ + 6e^- \rightarrow 2Cr^{3+} + 7H_2O$

Multiplying the iron(II) equation by 6 and adding gives:

$$Cr_2O_7^{2-}(aq) + 14H^+(aq) + 6Fe^{2+}(aq) \rightarrow 6Fe^{3+}(aq) + 2Cr^{3+}(aq) + 7H_2O(l)$$

The dichromate(VI) is orange before the reaction and the chromium(III) ion formed is green.

In the case of using acidifed manganate(VII), the overall equation is:

$$MnO_4^-(aq) + 8H^+(aq) + 5Fe^{2+}(aq) \rightarrow 5Fe^{3+}(aq) + Mn^{2+}(aq) + 4H_2O(l)$$

The manganate(VII) ion is intensely purple, whereas the aqueous manganese(II) ion is very pale pink and effectively colourless. Potassium manganate(VII) can therefore be used as a 'self-indicator' when carrying out titrations in which the amount of iron(II) ions is being measured. The end point is given by the first appearance of a permanent pink colour when $KMnO_4(aq)$ is added from a burette.

Examiners' tip

Don't try to remember overall redox equations — assemble them by building up the half-equations and then combine them.

Cobalt

Cobalt exists either in the common +2 state or in the rarer +3 state.

Oxidation state	+2	+3
Aqueous complex	$[Co(H_2O)_6]^{2+}$	$[Co(NH_3)_6]^{3+}$
Colour	Pink	Brown

Cobalt(II) ions can be converted into **cobalt(III)** ions using oxidising agents such as air in the presence of ammonia, or hydrogen peroxide in alkaline conditions.

In the presence of ammonia, the cobalt(II)–ammonia complex is oxidised by oxygen to form a dark brown solution containing the cobalt(III)–ammonia complex:

- oxidation of the cobalt(II)–ammonia complex:
  $[Co(NH_3)_6]^{2+} \rightarrow [Co(NH_3)_6]^{3+} + e^-$
- reduction of oxygen from the air:
  $O_2 + 2H_2O + 4e^- \rightarrow 4OH^-$

Multiplying the complex equation by 4 and adding gives:

$$4[Co(NH_3)_6]^{2+}(aq) + O_2(g) + 2H_2O(l) \rightarrow 4[Co(NH_3)_6]^{3+}(aq) + 4OH^-(aq)$$

With hydrogen peroxide in alkaline solution, cobalt(II) hydroxide is oxidised to form cobalt(III) hydroxide:

- oxidation of cobalt(II) hydroxide: $Co(OH)_2 + OH^- \rightarrow Co(OH)_3 + e^-$
- reduction of hydrogen peroxide: $H_2O_2 + 2e^- \rightarrow 2OH^-$

Multiplying the cobalt(II) equation by 2, adding and cancelling out OH^- gives:

$$2Co(OH)_2(s) + H_2O_2(aq) \rightarrow 2Co(OH)_3(s)$$

Calculations involving redox processes

Titration calculations involving manganate(VII) ions or dichromate(VI) ions are common in examinations. Both oxidising agents can oxidise many other ions (for example chloride, Cl^-, iron(II), Fe^{2+} and iodide, I^-). It is therefore a convenient way of determining the concentrations or amounts of these ions in a sample.

Example

In an experiment to determine the percentage by mass of iron in iron tablets, the following process was carried out.

Some iron tablets of total mass 3.67 g were ground up and dissolved in excess, warm, dilute sulfuric(VI) acid and the solution was then made up to 250 cm^3 in a volumetric flask. 25.0 cm^3 of this solution was removed by pipette and titrated against 0.0500 mol dm^{-3} potassium manganate(VII) solution. It was found that 18.50 cm^3 of the solution was required. Calculate the percentage by mass of iron in the tablets. [A_r of iron = 55.8]

Answer

$$MnO_4^- + 8H^+ + 5Fe^{2+} \rightarrow 5Fe^{3+} + Mn^{2+} + 4H_2O$$

Amount of MnO_4^- used $= \frac{18.50}{1000} \times 0.0500 = 9.25 \times 10^{-4}$ mol

Amount of Fe^{2+} reacting (from the equation) $= 9.25 \times 10^{-4}$ mol $\times 5$

$= 4.625 \times 10^{-3}$ mol

Total amount of Fe^{2+} in 250 cm^3 $= 4.625 \times 10^{-3} \times \frac{250}{25} = 0.04625$ mol

Mass of iron $= 0.09625 \times 55.8 = 2.58$ g

Percentage by mass of iron $= \frac{2.58}{3.67} \times 100$

$= 70.3\%$

Now test yourself Tested

4 In an experiment to determine the quantity of iodide in a sample of dried seaweed, 50.0 g of the seaweed was treated in water and the released iodide ions were made up to 250 cm^3 in a volumetric flask. 25.0 cm^3 of the solution was removed, an indicator added and the solution was found to react with 21.25 cm^3 of 0.0100 mol dm^{-3} potassium dichromate(VI) solution.

Calculate the percentage of iodine in the seaweed. [A_r of iodine = 126.9]

Answers on p. 111

Catalytic behaviour of transition metals

Revised

There are many **catalysts** but only two types of **catalysis** involving transition metals — **homogeneous** and **heterogeneous**.

A **catalyst** is a substance that increases the rate of a chemical reaction by providing an alternative route for the reaction with a lower activation energy. At the end of the reaction, the catalyst is chemically unchanged.

Homogeneous catalysis

In **homogenous catalysis**, the catalyst usually forms an **intermediate species** during the reaction. Typically a transition metal ion can change its oxidation state from a higher to a lower one, and then back up to a higher state (or vice versa). The ability of transition metals to change their oxidation state in this way makes them very efficient catalysts for redox reactions.

Homogeneous catalysis involves a catalyst in the same physical state as the reactants.

In the reaction between peroxodisulfate(VI) ions and iodide ions:

$$S_2O_8^{2-}(aq) + 2I^-(aq) \rightarrow 2SO_4^{2-}(aq) + I_2(aq)$$

This reaction is normally slow because two negative ions need to interact. If iron(II) ions are added, the following processes takes place more quickly because oppositely charged ions interact in each stage.

Electrode potentials are often useful in predicting what may happen in each stage:

$Fe^{3+}(aq) + e^- \rightleftharpoons Fe^{2+}(aq)$ $E^\ominus = +0.77\,V$

$I_2(aq) + 2e^- \rightleftharpoons 2I^-(aq)$ $E^\ominus = +0.54\,V$

$S_2O_8^{2-}(aq) + 2e^- \rightleftharpoons 2SO_4^{2-}(aq)$ $E^\ominus = +2.01\,V$

Step 1: $S_2O_8^{2-}$ ions react with Fe^{2+} ions:

$S_2O_8^{2-}(aq) + 2e^- \rightarrow 2SO_4^{2-}(aq)$ $E^\ominus = +2.01\,V$

$Fe^{2+}(aq) \rightarrow Fe^{3+}(aq) + e^-$ $E^\ominus = -0.77\,V$

Adding gives $2Fe^{2+}(aq) + S_2O_8^{2-}(aq) \rightarrow 2SO_4^{2-}(aq) + 2Fe^{3+}(aq)$

where $E^\ominus = +2.01 + (-0.77) = +1.24\,V$

Step 2: Fe^{3+} ions react with I^- ions:

$Fe^{3+}(aq) + e^- \rightarrow Fe^{2+}(aq)$ $E^\ominus = +0.77\,V$

$2I^-(aq) \rightarrow I_2(aq) + 2e^-$ $E^\ominus = -0.54\,V$

Adding gives $2Fe^{3+}(aq) + 2I^-(aq) \rightarrow I_2(aq) + 2Fe^{2+}(aq)$

where $E^\ominus = +0.77 + (-0.54) = +0.23\,V$

Notice how iron(II) is consumed in the first step, and then reformed in the second step. Either Step 1 or Step 2 can occur first. Fe^{3+} can also act as the catalyst.

Another reaction in which the catalyst is in the same phase as the reactants is that between ethanedioate ions and manganate(VII) ions:

$$2MnO_4^-(aq) + 16H^+(aq) + 5C_2O_4^{2-}(aq) \rightarrow 2Mn^{2+}(aq) + 8H_2O(l) + 5CO_2(aq)$$

This reaction is also very slow under normal conditions because two negative ions need to react. The reaction is catalysed by manganese(II) ions. Because these ions are formed in the reaction itself, the process is known as **autocatalysis**.

Heterogeneous catalysis

In **heterogenous catalysis**, the catalyst is in a phase that is different from that of the reactants.

Heterogeneous catalysis involves a catalyst in a different physical state from the reactants.

Examples of heterogeneous catalysis include:

- using iron in the Haber Process
  $N_2(g) + 3H_2(g) \rightleftharpoons 2NH_3(g)$
- using chromium(III) oxide in manufacture of methanol
  $CO(g) + 2H_2(g) \rightarrow CH_3OH(l)$
- using rhodium, palladium and platinum in catalytic converters in car exhausts
  $2CO(g) + 2NO(g) \rightarrow CO_2(g) + N_2(g)$
- using vanadium(V) oxide in the Contact process
  $2SO_2(g) + O_2(g) \rightleftharpoons 2SO_3(g)$

In this type of catalysis, the reacting molecules are **adsorbed** onto the surface of the catalyst. The reaction then takes place on the surface. Clearly, maximising the surface area by providing a support medium will increase the rate of reaction further and reduce potential costs. For example, rhodium is used on a ceramic support in catalytic converters in cars.

Sometimes, for example in the Haber Process and in catalytic converters, the catalyst may become **poisoned**. Foreign atoms bond to the active sites on the surface of the catalyst reducing the effective surface area available for reaction. This means that there will be costs incurred due to the inefficient operation of catalysts.

In the Contact process, vanadium(V) oxide provides a surface on which the reaction takes place, but it also reacts by changing the oxidation state of the vanadium forming an intermediate, V_2O_4:

$$V_2O_5(s) + SO_2(g) \rightarrow V_2O_4(s) + SO_3(g)$$
$$V_2O_4(s) + \tfrac{1}{2}O_2(g) \rightarrow V_2O_5(s)$$
$$\text{Overall: } SO_2(g) + \tfrac{1}{2}O_2(g) \rightarrow SO_3(g)$$

Reactions of hydrated transition metal ions

When ligands bond to transition metal ions, the ligands act as **Lewis bases** — they donate lone pairs to form co-ordinate bonds. Transition metal ions accept lone pairs and are therefore **Lewis acids**.

A **Lewis acid** is a lone-pair acceptor. A **Lewis base** is a lone-pair donor.

Figure 13.10 shows a typical hydrated complex involving six water molecules forming co-ordinate bonds to a central transition metal ion, M^{n+}, to form an octahedral complex.

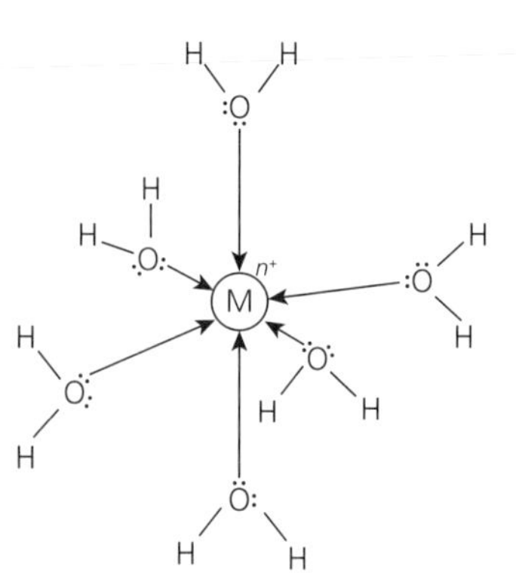

Figure 13.10 Typical octahedral aqua-complex

Aqueous ions in solution

Revised

Many transition metals form aqueous ions with charges of +2 or +3:

+2 ions: $[Fe(H_2O)_6]^{2+}$ $[Co(H_2O)_6]^{2+}$ $[Cu(H_2O)_6]^{2+}$

+3 ions: $[Fe(H_2O)_6]^{3+}$ $[Co(H_2O)_6]^{3+}$ $[Cr(H_2O)_6]^{3+}$

When these ions are in solution, equilibria exist in which the complex ion undergoes a hydrolysis process (effectively the loss of a proton, H^+). For example:

$$[Fe(H_2O)_6]^{2+} + H_2O \rightleftharpoons [Fe(H_2O)_5(OH)]^{+} + H_3O^{+}$$
$$[Fe(H_2O)_6]^{3+} + H_2O \rightleftharpoons [Fe(H_2O)_5(OH)]^{2+} + H_3O^{+}$$

- A +3 metal ion is smaller and more highly charged than a +2 ion, so it has a higher charge density.
- A +3 metal ion polarises water molecules to a greater extent than a +2 ion.
- This results in bond weakening of O–H bonds in the +3 aqueous complex.
- As a result, +3 aqueous ions are more acidic than +2 ions.

So, the equilibrium relating to the +3 ion has an equilibrium position that lies more to the right-hand side than the corresponding +2 equilibrium position.

As a result, a solution of +3 aqueous ions has a lower pH than a solution of similar concentration of +2 ions.

Transition aqueous metal ions in the +2 oxidation state

- All that form M^{2+} ions have +2 as the most common oxidation state (by loss of both s electrons).
- All the aqueous ions have the form of octahedral $[M(H_2O)_6]^{2+}$.
- Some common metal(II) aqueous ions colours are:

Fe^{2+}	Co^{2+}	Cu^{2+}
pale green	pink	blue

Reactions of +2 aqueous ions

With sodium hydroxide solution

All aqueous ions react with hydroxide ions and form insoluble and coloured metal(II) hydroxides as precipitates.

Aqueous ion	Fe^{2+}	Co^{2+}	Cu^{2+}
Name of precipitate	Iron(II) hydroxide	Cobalt(II) hydroxide	Copper(II) hydroxide
Formula of precipitate	$Fe(OH)_2$	$Co(OH)_2$	$Cu(OH)_2$
Colour of precipitate	Dark green	Blue/green	Light blue

The reaction taking place is between hydroxide ions, OH^-, and the aqueous transition metal complex:

$$[M(H_2O)_6]^{2+}(aq) + 2OH^-(aq) \rightarrow [M(H_2O)_4(OH)_2](s) + 2H_2O(l)$$

or simply:

$$M^{2+}(aq) + 2OH^-(aq) \rightarrow M(OH)_2(s)$$

Ammonia solution

Ammonia is a weak base in aqueous solution and contains hydroxide ions, OH^-:

$$NH_3(aq) + H_2O(aq) \rightleftharpoons NH_4^+(aq) + OH^-(aq)$$

The hydroxide ions remove protons from the aqueous +2 complex in the first step:

$$[M(H_2O)_6]^{2+}(aq) + 2OH^-(aq) \rightarrow [M(H_2O)_4(OH)_2](s) + 2H_2O(l)$$

In some cases, the hydroxide precipitate dissolves in excess ammonia solution. In this process, ammonia, NH_3, acts as a ligand and removes the water molecules and hydroxide ions from the original complex in a process known as **ligand substitution**:

$$[M(H_2O)_4(OH)_2](s) + 6NH_3(aq) \rightleftharpoons [M(NH_3)_6]^{2+}(aq) + 4H_2O(l) + 2OH^-(aq)$$

- Iron(II) aqueous ions form iron(II) hydroxide in the first stage, but no further ligand substitution takes place.
- Cobalt(II) aqueous ions form a dark blue precipitate of cobalt(II) hydroxide; this dissolves in excess ammonia to form the straw-coloured cobalt(II)–ammonia complex $[Co(NH_3)_6]^{2+}$.
- Copper(II) aqueous ions form a light blue precipitate of copper(II) hydroxide; this dissolves in excess ammonia to form the dark blue complex $[Cu(NH_3)_4(H_2O)_2]^{2+}$.

Sodium carbonate solution

The carbonate ions, CO_3^{2-}, react with aqueous 2+ ions to form insoluble carbonate precipitates:

$$M^{2+}(aq) + CO_3^{2-}(aq) \rightarrow MCO_3(s)$$

Name of precipitate	Iron(II) carbonate	Cobalt(II) carbonate	Copper(II) carbonate
Formula of precipitate	$FeCO_3$	$CoCO_3$	$CuCO_3$
Colour of precipitate	Green	Pink	Green-blue

Transition metal ions in the +3 oxidation state

In aqueous solution, they all exist as the $[M(H_2O)_6]^{3+}$ ion although **hydrolysis** always occurs to form the $[M(H_2O)_5(OH)]^{2+}$ ion along with protons — so solutions containing these ions are weak acids. Aqueous solutions of Cr^{3+}, Fe^{3+} and Al^{3+} (for comparison) tend to be acidic and have pK_a values similar to that of ethanoic acid.

Reactions of aqueous +3 ions

Sodium hydroxide solution

The equilibrium written below shifts to the right-hand side on adding hydroxide ions, OH^-, by reacting with hydrogen ions:

$$[Fe(H_2O)_6]^{3+}(aq) + H_2O(l) \rightleftharpoons [Fe(H_2O)_5(OH)]^{2+}(aq) + H_3O^+(aq)$$

or

$$[Fe(H_2O)_6]^{3+}(aq) + OH^-(aq) \rightarrow [Fe(H_2O)_5(OH)]^{2+}(aq) + H_2O(l)$$

This process repeats itself until the metal(III) hydroxide precipitate (with zero charge) forms.

- Iron(III) aqueous ions form a brown precipitate of hydrated iron(III) hydroxide:

  $$[Fe(H_2O)_6]^{3+}(aq) + 3OH^-(aq) \rightarrow [Fe(OH)_3(H_2O)_3](s) + 3H_2O(l)$$

- Chromium(III) aqueous ions form a green precipitate of chromium(III) hydroxide, but this dissolves in excess hydroxide ions to form a green solution containing $[Cr(OH)_6]^{3-}$ ions:

  $$[Cr(H_2O)_6]^{3+}(aq) + 3OH^-(aq) \rightarrow [Cr(OH)_3(H_2O)_3](s) + 3H_2O(l)$$

  then

  $$[Cr(OH)_3(H_2O)_3](s) + 3OH^-(aq) \rightarrow [Cr(OH)_6]^{3-}(aq) + 3H_2O(l)$$

- Aluminium aqueous ions have a similar reaction to that of chromium(III) ions, although the hydroxide precipitate is white and a colourless solution forms on adding excess sodium hydroxide solution:

  $$[Al(H_2O)_6]^{3+}(aq) + 3OH^-(aq) \rightarrow [Al(OH)_3(H_2O)_3](s) + 3H_2O(l)$$

  then

  $$[Al(OH)_3(H_2O)_3](s) + OH^-(aq) \rightarrow [Al(OH)_4(H_2O)_2]^-(aq) + H_2O(l)$$

  The hydroxides of aluminium and chromium(III) are **amphoteric** because they react with both acids and bases to form salts.

Examiners' tip

Many of the reactions of +3 aqueous ion solutions are those that an acid would undergo, particularly when reacting with carbonates to form carbon dioxide gas.

Ammonia solution

Ammonia is a weak base in aqueous solution and contains hydroxide ions, OH^-:

$$NH_3(aq) + H_2O(aq) \rightleftharpoons NH_4^+(aq) + OH^-(aq)$$

The hydroxide ions form the metal(III) hydroxides as precipitates — the same as those produced using sodium hydroxide solution.

Name of precipitate	Iron(III) hydroxide	Chromium(III) hydroxide	Aluminium hydroxide
Formula of precipitate	$Fe(OH)_3$	$Cr(OH)_3$	$Al(OH)_3$
Colour of precipitate	Brown	Green	White

When excess ammonia solution is added:

- iron(III) hydroxide precipitate is unaffected
- chromium(III) hydroxide dissolves slowly and forms the violet complex ion $[Cr(NH_3)_6]^{3+}$
- aluminium hydroxide precipitate is unaffected

Sodium carbonate solution

An aqueous solution of a carbonate is alkaline:

$$CO_3^{2-}(aq) + H_2O(l) \rightleftharpoons HCO_3^-(aq) + OH^-(aq)$$

So a reaction is expected in which metal(III) hydroxide precipitates form.

However, carbonate ions will also react with the acidic aqueous metal(III) ions to form carbon dioxide gas:

$$CO_3^{2-}(aq) + H_3O^+(aq) \rightarrow CO_2(g) + H_2O(l)$$

In the case of chromium(III) ions in solution, this reaction takes place:

$$2[Cr(H_2O)_6]^{3+}(aq) + 3CO_3^{2-}(aq) \rightarrow 2[Cr(H_2O)_3(OH)_3](s) + 3CO_2(g) + 3H_2O(l)$$

$[Fe(H_2O)_6]^{3+}(aq)$ and $[Al(H_2O)_6]^{3+}(aq)$ react in the same way as $[Cr(H_2O)_6]^{3+}(aq)$.

Examiners' tip

Remember that aqueous solutions containing +3 metal ions are acidic. They will react with sodium carbonate to form carbon dioxide gas and so fizzing will be seen.

Typical mistake

Do not make an error by stating that a metal(III) carbonate forms when a carbonate is added to an aqueous metal(III) ion solution. The aqueous ion is too acidic and decomposes the carbonate ion to form carbon dioxide gas — so a metal(III) hydroxide forms instead.

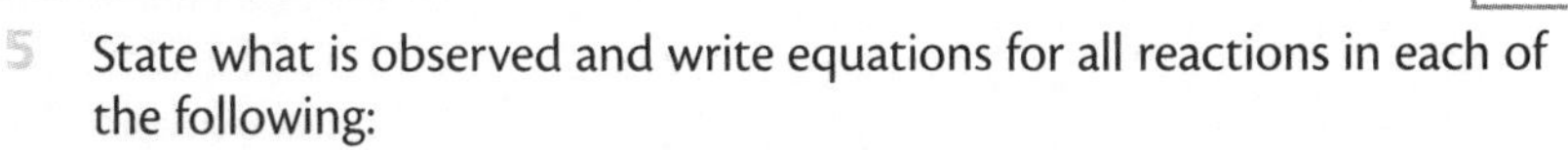

5 State what is observed and write equations for all reactions in each of the following:
 (a) ammonia solution is added dropwise until in excess to aqueous copper(II) ions
 (b) sodium hydroxide solution is added dropwise until in excess to chromium(III) ions
 (c) sodium carbonate solution is added to aqueous aluminium(III) sulfate solution.

6 Explain how sodium carbonate solution can be used to distinguish between a solution containing iron(II) ions and one containing iron(III) ions.

Answers on p. 111

Ligand substitution reactions

Ligand substitution is a process that can happen to a complex ion depending on conditions.

Ligand substitution reactions happen when one ligand in a complex is replaced by another.

Ligands which are molecules of similar size — for example ammonia and water — may be interchanged without any change of co-ordination number:

$$[Co(H_2O)_6]^{2+}(aq) + 6NH_3(aq) \rightleftharpoons [Co(NH_3)_6]^{2+}(aq) + 6H_2O(l)$$

$$[Cr(H_2O)_6]^{3+}(aq) + 6NH_3(aq) \rightleftharpoons [Cr(NH_3)_6]^{3+}(aq) + 6H_2O(l)$$

You can see that the co-ordination number of all the complexes remains at 6.

As mentioned in an earlier section, aqueous copper(II) ions react with ammonia ligands, but the substitution is not complete — only four ammonia ligands bond to the copper(II) ion instead of six:

$$[Cu(H_2O)_6]^{2+}(aq) + 4NH_3(aq) \rightleftharpoons [Cu(NH_3)_4(H_2O)_2]^{2+}(aq) + 4H_2O(l)$$

So ligand substitution has happened but the co-ordination number is still 6.

Ligands of period 3 elements, such as chloride ions (Cl^-), are larger than those of period 2 elements (e.g. F^-, H_2O, NH_3). This means that fewer of the larger ligands can bond to the metal, so the co-ordination number decreases — from 6 to 4 in the examples below:

$$[Cu(H_2O)_6]^{2+}(aq) + 4Cl^-(aq) \rightleftharpoons [CuCl_4]^{2-}(aq) + 6H_2O(l)$$

$$[Co(H_2O)_6]^{2+}(aq) + 4Cl^-(aq) \rightleftharpoons [CoCl_4]^{2-}(aq) + 6H_2O(l)$$

When multidentate ligands are used — for example ethane-1,2-diamine, ethanedioate ions (both bidentate ligands) or EDTA (hexadentate) — then stable complexes are formed:

$$[Cu(H_2O)_6]^{2+}(aq) + EDTA^{4-}(aq) \rightleftharpoons [Cu(EDTA)]^{2-}(aq) + 6H_2O(l)$$

The **chelating effect** of using a ligand like EDTA results in a highly stable complex. One of the reasons is that there is a large increase in entropy when the reaction takes place, due to many water molecules being displaced:

$$[Cr(H_2O)_6]^{2+}(aq) + EDTA^{4-}(aq) \rightleftharpoons [Cr(EDTA)]^{2-}(aq) + 6H_2O(l)$$

There are 2 species on the left-hand side of the equilibrium and 7 species on the right-hand side. There will therefore be an increase in entropy of the system. This will decrease the value for the Gibbs free energy, and hence make the complex more thermodynamically stable.

Now test yourself Tested

7 This ligand is known by the abbreviation 'TPEDA'.

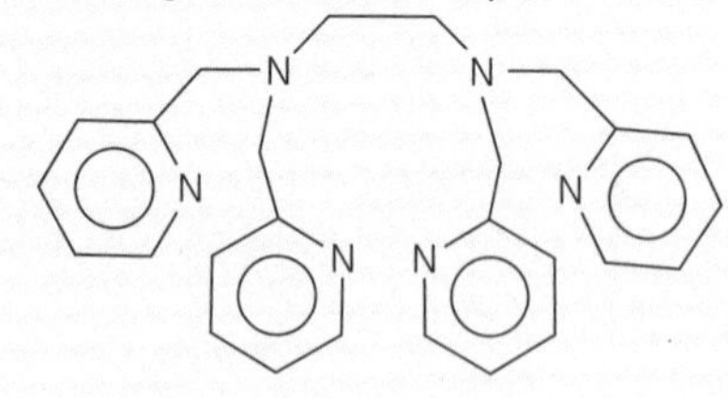

(a) Suggest how many co-ordinate bonds TPEDA will form with a transition metal ion.

(b) Write an equation to show how TPEDA would react with aqueous iron(II) ions.

(c) Suggest why the iron(II)–TPEDA complex is thermodynamically stable.

Answers on p. 111

Exam practice

1 **(a)** In terms of bonding, explain, the term 'complex'. [2]

(b) Identify **one** of the species in this list that does **not** act as a ligand. Explain your answer. [2]

CN^- O^{2-} CH_4 NH_3

(c) The element gold is in the d-block of the periodic table. Consider the following gold complex (known as 'audien'), which contains a chloride ion, Cl^-.

$$\left[\begin{array}{c} \text{NH (bridged to both } NH_2 \text{ groups)} \\ | \\ NH_2 - Au - NH_2 \\ | \\ Cl \end{array}\right]^{2+}$$

(i) Deduce the oxidation state of gold in this complex ion. [1]

(ii) Give the co-ordination number of gold in this complex ion. [1]

(iii) Give the names of two possible shapes around the gold atom in audien. [2]

2 This question is about cobalt chemistry.

(a) Aqueous cobalt(II) ions, $[Co(H_2O)_6]^{2+}(aq)$, are pink.

(i) With reference to electrons, explain why aqueous cobalt (II) ions are pink. [3]

(ii) By reference to aqueous cobalt (II) ions, state the meaning of each of the terms in the equation $\Delta E = h\nu$. [3]

(iii) Write an equation for the reaction, in aqueous solution, between $[Co(H_2O)_6]^{2+}$ and an excess of chloride ions. [2]

(iv) State the shape of the complex produced in part (a)(iii) and explain why its shape differs from that of the $[Co(H_2O)_6]^{2+}$ ion. [3]

(v) Draw the structure of the ethanedioate ion ($C_2O_4^{2-}$). Explain how this ion can act as a ligand. [2]

(b) When a dilute solution containing ethanedioate ions is added to a solution containing cobalt(II) ions, a substitution reaction occurs. In this reaction, four water molecules are replaced and a new complex is formed.

(i) Write an ionic equation for the reaction. Give the co-ordination number of the complex formed and name its shape. [4]

(ii) In the complex formed, the two water molecules are opposite each other. Draw a diagram to show how the ethanedioate ions are bonded to the cobalt(II) ion and give a value for the O–Co–O bond angles. Don't show the water molecules in your diagram. [2]

Answers and quick quizzes online

Online

Examiners' summary

You should now have an understanding of:

- the general properties of transition metals
- complexes
- ligands and the different types that exist
- the co-ordination number of a complex
- the various shapes of complexes
- how colour arises in a complex
- the variable oxidation states of transition metals
- how to interconvert different oxidation states involving chromium, iron and cobalt
- catalysis in terms of heterogeneous and homogeneous processes
- how homogeneous catalysis works
- how heterogeneous catalysis works and how catalysts may be poisoned
- various applications of transition metal catalysts
- Lewis acids and bases
- the existence of transition metal +2 and +3 aqueous ions
- the relative acidity of +2 and +3 ions in terms of the charge density of the positive ion
- the reactions of various +2 and +3 ions with ammonia, hydroxide ions and carbonate ions
- ligand substitution reactions and the factors that affect the co-ordination numbers in different complexes
- how the thermodynamic stability of complexes can be related to entropy effects during ligand substitution

Now test yourself answers

Chapter 1

1 **(a)** Order with respect to $(CH_3)_3CBr$ = 1; order with respect to OH^- = 0

(b) Overall order = 1 + 0 = 1

(c) $\frac{\text{mol dm}^{-3}\,\text{s}^{-1}}{\text{mol dm}^{-3}} = \text{s}^{-1}$

2 **(a)** Order with respect to NO = 2; order with respect to H_2 = 1

(b) Overall order = 2 + 1 = 3

(c) Rate = $k[NO]^2[H_2]^1$; or simplified as = $k[NO]^2[H_2]$

(d) Rate = $k[NO]^2[H_2]^1$; substituting data from experiment 1

$1.11 \times 10^{-3} = k(0.100)^2(0.100)$; $k = 1.11\,\text{mol}^{-2}\,\text{dm}^6\,\text{s}^{-1}$

Chapter 2

1 **(a)** $K_c = \frac{[HI(g)]^2}{[H_2(g)][I_2(g)]}$

(b) $K_c = \frac{[PCl_3(g)][Cl_2(g)]}{[PCl_5(g)]}$

(c) No units and mol dm^{-3} respectively

2

	$PCl_5(g) \rightleftharpoons$	$PCl_3(g)$ +	$Cl_2(g)$
Start amounts/mol	0.550	0	0
PCl_5 decreases to 0.240 mol (down by 0.310 mol)			
Equilibrium amounts/mol	0.240	0.310	0.310
Equilibrium conc/mol dm^{-3}	0.120	0.155	0.155

$$K_c = \frac{[PCl_3(g)][Cl_2(g)]}{[PCl_5(g)]} = \frac{0.155^2}{0.120} = 0.200\,\text{mol dm}^{-3}$$

3 **(a)** $K_c = \frac{[B(aq)][C(aq)]}{[A(aq)]}$

(b) The forward reaction is endothermic; an increase in temperature favours the forward direction; as K_c increases

(c) No change; the catalyst affects only the rate of reaction and not the equilibrium position; the rates of the forward and reverse reactions are increased equally

Chapter 3

1 HNO_3 is base 1; $H_2NO_3^+$ is conjugate acid 1

H_2SO_4 is acid 2; HSO_4^- is conjugate base 2

2 **(a)** $C_6H_5COOH(aq) + H_2O(l) \rightleftharpoons C_6H_5COO^-(aq) + H_3O^+(aq)$

(b) $CH_3NH_2(aq) + H_2O(l) \rightleftharpoons CH_3NH_3^+(aq) + OH^-(aq)$

3 **(a)** $pH = -\log_{10}[H_3O^+] = -\log_{10}(6.60 \times 10^{-2}) = 1.18$

(b) $K_w = [H_3O^+][OH^-(aq)]$

$$[H_3O^+] = \frac{1.00 \times 10^{-14}}{5.67 \times 10^{-4}} = 1.76 \times 10^{-11}\,\text{mol dm}^{-3}$$

pH = 10.75

(c) $K_w = [H_3O^+(aq)][OH^-(aq)]$

$$[H_3O^+] = \frac{1.00 \times 10^{-14}}{0.0500} = 2.00 \times 10^{-13}\,\text{mol dm}^{-3}$$

pH = 12.70

(d) $pH = -\log_{10}[H_3O^+] = -\log_{10}(2 \times 0.0950) = 0.72$

4 $[H_3O^+] = 10^{-pH} = 10^{-6.45} = 3.55 \times 10^{-7}\,\text{mol dm}^{-3}$

$K_w = [H_3O^+(aq)]\,[OH^-(aq)]$

$$[OH^-] = \frac{1.00 \times 10^{-14}}{3.55 \times 10^{-7}} = 2.82 \times 10^{-8}\,\text{mol dm}^{-3}$$

5 **(a)** $CH_3COOH(aq) + H_2O(l) \rightleftharpoons CH_3COO^-(aq) + H_3O^+(aq)$

(b) $K_a = \frac{[CH_3COO^-(aq)][H_3O^+(aq)]}{[CH_3COOH(aq)]}$

(c) $[H_3O^+] = \sqrt{K_a \times [HA]} = \sqrt{(1.74 \times 10^{-5}) \times 0.100}$
$= 1.32 \times 10^{-4}\,\text{mol dm}^{-3}$

$pH = -\log_{10}(1.32 \times 10^{-4}) = 2.88$

Chapter 4

1 **(a)** Molecules with the same molecular formula but different structures

(b)

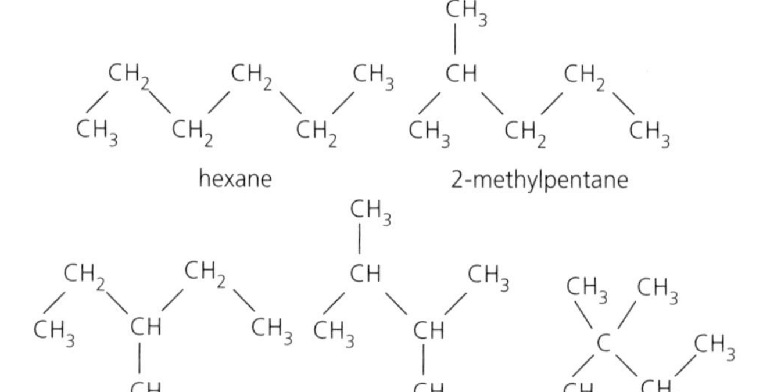

2 **(a)** $C_{16}H_{30}O_2$

(b) It has two hydrogen atoms on each side on the carbon–carbon double bond that can be either on the same side (Z) of the double bond or on opposite sides (E); lack of free rotation about the carbon–carbon double bond makes this possible

(c)

H, $(CH_2)_7COOH$ / C=C / $CH_3(CH_2)_5$, H

3 **(a)** Molecules having the same structural formula but different spatial arrangements of atoms

(b)

COOH, C, H, OH, H_3C | COOH, C, HO, H, CH_3

(c) One enantiomer will rotate the plane of polarised light in one direction; the other enantiomer (at equal concentration) will rotate the plane of plane-polarised light equally in the other direction.

Chapter 5

1 Aldehydes: cinnamaldehyde and vanillin; Ketones: carvone and progesterone

2 **(a)**

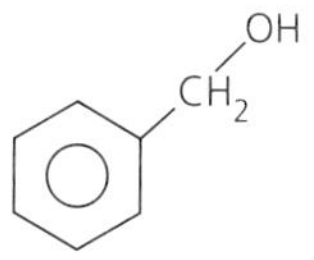

(b)

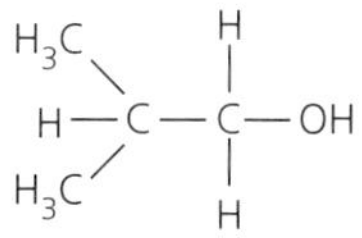

(c)

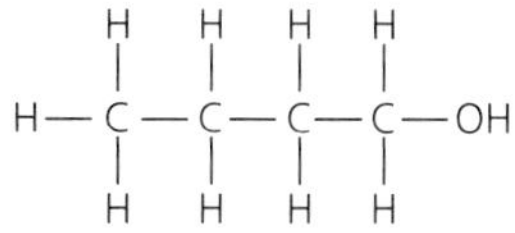

3

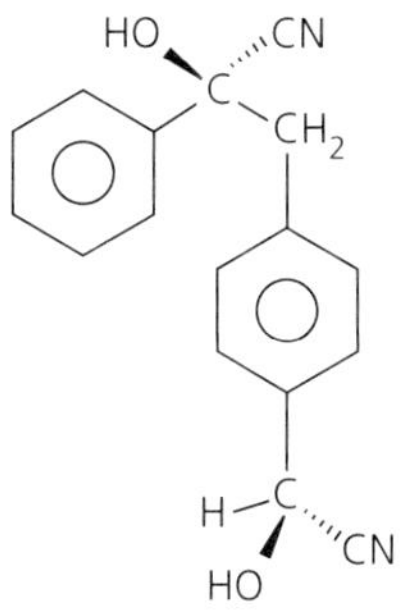

4 **(a)**

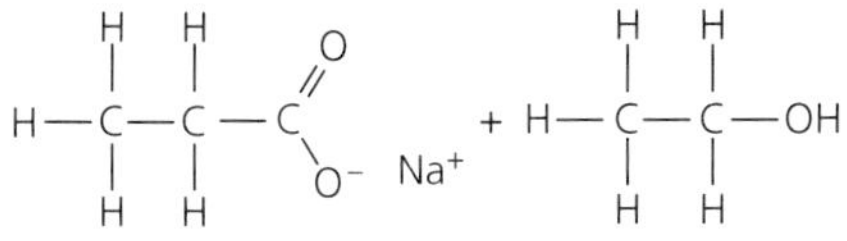

(b)

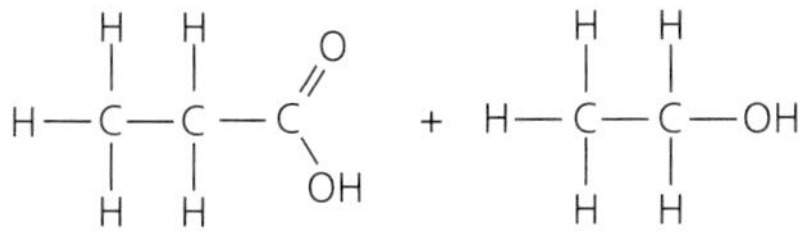

5

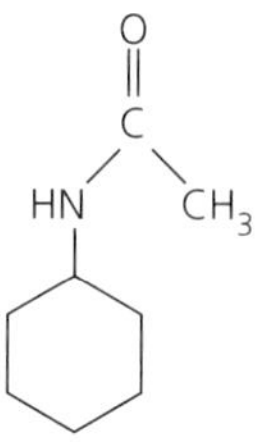

Chapter 6

1 **(a)** Reaction 1: N^+O_2

Reaction 2: CH_3C^+O

(b) Reaction 1: concentrated nitric(V) acid and concentrated sulfuric(VI) acid; maximum temperature 50°C

Reaction 2: CH_3COCl; $AlCl_3$ or $FeCl_3$

Chapter 7

1 **(a)**

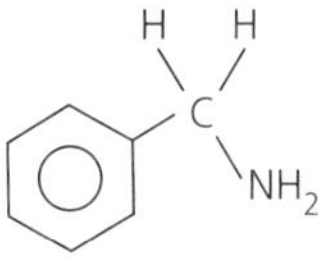

(b)

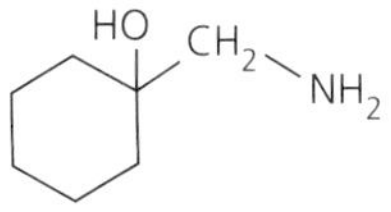

2 **(a)**

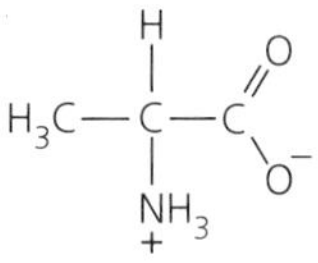

(b)

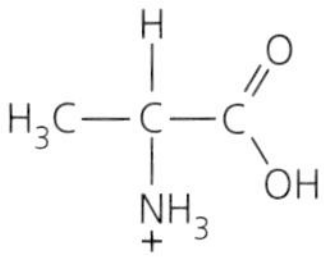

(c)

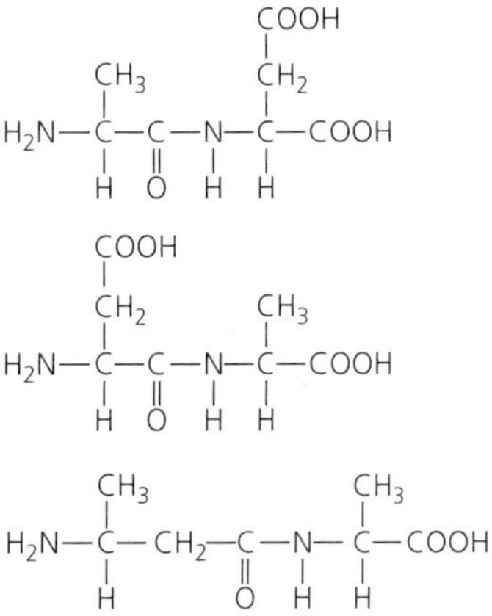

Chapter 8

1 **(a)**

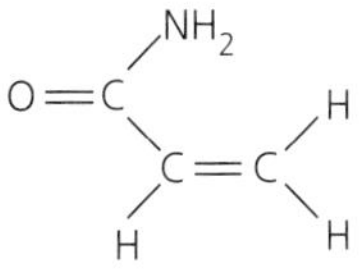

(b)

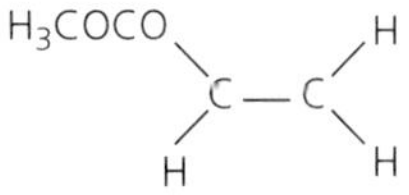

2

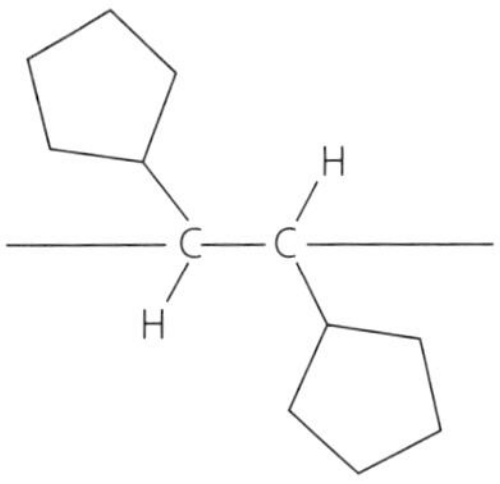

3 HO–C(=O)–(cyclohexane-1,3-diyl)–C(=O)–OH and HO–CH_2–(cyclohexane-1,3-diyl)–CH_2–OH

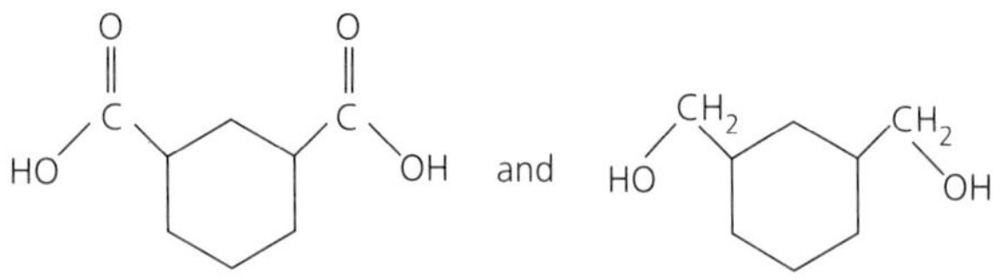

4 **(a)** An amino acid

(b)

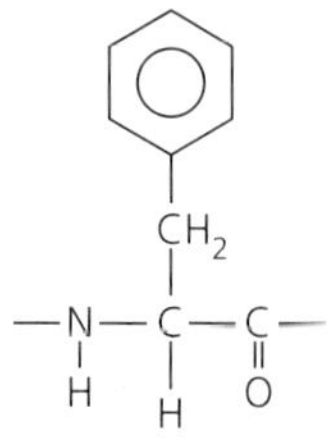

—N—C—C—
H | CH_2 (phenyl) | H | O

(c)

HN, C=O, NH, CH, CH_2 (phenyl), CH, CH_2 (phenyl), O=C

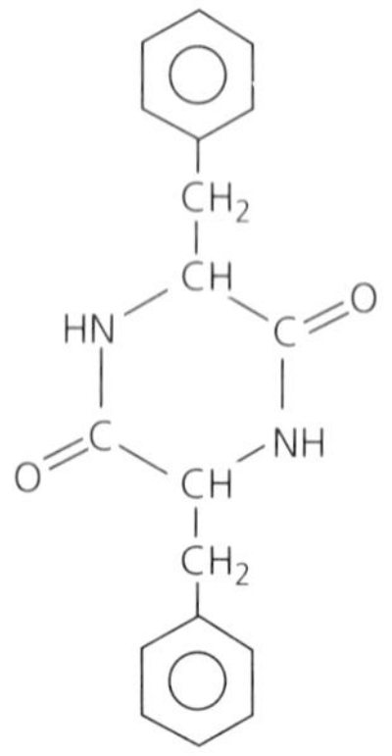

Chapter 9

1 **(a)** OH alcohols; a stretch occurs at about 3350 cm^{-1}

(b) $M_r = 60$

(c) 26.7% oxygen

Amounts in mol: $\frac{60}{12.0}$ = 5.0 mol carbon;

$\frac{13.3}{1.0}$ = 13.3 mol hydrogen;

$\frac{26.7}{16.0}$ = 1.67 mol oxygen;

Simplest ratio is 3:8:1 or C_3H_8O

(d) C_3H_8O

(e) X must have an OH present and contain 3 carbon atoms; so could be either propan-1-ol or propan-2-ol

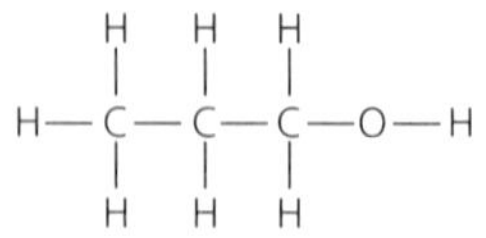

H–C(H)(H)–C(H)(H)–C(H)(H)–O–H

propan-1-ol

H–C(H)(H)–C(H)(O–H)–C(H)(H)–H

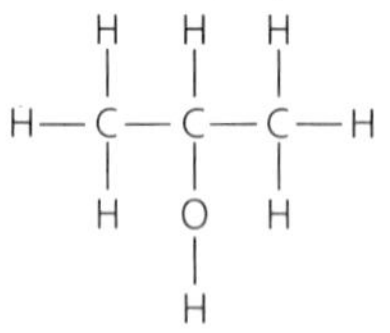

propan-2-ol

2 The molecule has three proton environments that will generate areas in the ratio 1:2:1 (moving from left to right along the molecule). The left-hand CHO group will give a peak at δ 10–11; the right-hand CHO group will give peak at a slightly different δ value; the central CH_2 group would give a peak at a higher δ of 5–6.

The peak from the CHO group on the left would be a singlet. The CH_2 protons peak would be split into a doublet by the proton in the right-hand CHO group. The peak from the CHO group on the right would be split into a triplet by the protons in the CH_2 group.

Chapter 10

1 Polarising cations that have a high charge density induce more covalent character in their compounds by polarising or distorting the spherical anion. Therefore both calcium compounds are likely to be more covalent; particularly if combined with a large negative ion like the sulfide ion, S^{2-}. So, the correct answer is **A**.

2 Lattice enthalpy depends on the same factors as hydration energy — ionic charge and ionic radius. The substance with the most exothermic lattice formation energy will be the one consisting of small, highly charged positive ions (a high charge density) and also negative ions that have a high charge and small size. The correct answer is **B**.

3 Charge and size govern how strongly ions are attracted to water molecules — a high charge and small ionic radius are favourable for strong attraction. The calcium ion is the only one with a double charge. Hydration energy is considerably more affected by charge than ionic radius, so **B** will be the most exothermic.

4 **(a)** $\Delta H = 944 + (2 \times +436) - [145 + (4 \times +388)] = 1816 - 1697 = +119\,kJ\,mol^{-1}$

(b) $\Delta H = 157 + (2 \times +463) - (436 + 496) = 1083 - 932 = +151\,kJ\,mol^{-1}$

(c) $\Delta H = 157 + (2 \times +463) - [(2 \times +463) + (½ \times + 496)] = 1083 - 1174 = -91\,kJ\,mol^{-1}$

5 The enthalpy of formation of hydrazine assumes the formation of liquid hydrazine, not gaseous hydrazine; mean bond energies were used in calculating ΔH in Q4 part (a) rather than specific values

6 **(a)** Decrease

(b) Increase

(c) Increase

(d) Decrease

(e) Decrease

7 **(a)** $\Delta S^{\ominus} = \Sigma S^{\ominus}$ (products) $- \Sigma S^{\ominus}$ (reactants)

$= + 93 - [40 + 214] = 93 - 254 = -161\,J\,K^{-1}\,mol^{-1}$

(b) $CO_2(g)$ is being used up in the reaction; the number of moles of gas decreases; and more order will form in the reaction; or there will be a decrease in disorder

Chapter 11

1 **(a)** $4Al(s) + 3O_2(g) \rightarrow 2Al_2O_3(s)$

(b) $S(s) + O_2(g) \rightarrow SO_2(g)$

(c) $4Na(s) + O_2(g) \rightarrow 2Na_2O(s)$

2 $2Cs(s) + 2H_2O(l) \rightarrow 2CsOH(aq) + H_2(g)$; the likely pH of the solution formed will be 13–14; caesium hydroxide is a strong alkali

3 **(a)** $SO_2(g) + H_2O(l) \rightarrow H_2SO_3(aq)$

(b) $Na_2O(s) + H_2O(l) \rightarrow 2NaOH(aq)$

4 $2NaOH(aq) + H_2SO_3(aq) \rightarrow Na_2SO_3(aq) + 2H_2O(l)$

Chapter 12

1 **(a)** $Cr_2O_7^{2-}(aq) + 14H^+(aq) + 6e^- \rightarrow 2Cr^{3+}(aq) + 7H_2O(l)$

$2I^-(aq) \rightarrow I_2(aq) + 2e^-$

(b) (i) Chromium(VI) in dichromate(VI)

(ii) Iodide ions

(c) $Cr_2O_7^{2-}(aq) + 14H^+(aq) + 6I^-(aq) \rightarrow 2Cr^{3+}(aq) + 3I_2(aq) + 7H_2O(l)$

(d) (i) +6 to +3

(ii) −1 to 0

2 **(a)** $FeO_4^{2-}(aq) + 8H^+(aq) + 3e^- \rightarrow Fe^{3+}(aq) + 4H_2O(l)$

$MnO_2(s) + 2H_2O(l) \rightarrow MnO_4^-(aq) + 4H^+(aq) + 3e^-$

(b) (i) Iron(VI) in ferrate(VI)

(ii) Manganese(IV) in MnO_2

(c) $FeO_4^{2-}(aq) + 4H^+(aq) + MnO_2(s) \rightarrow Fe^{3+}(aq) + 2H_2O(l) + MnO_4^-(aq)$

(d) (i) +4 to +7

(ii) +6 to +3

3 $Fe^{3+}(aq) + e^- \rightleftharpoons Fe^{2+}(aq)$; $E^\ominus = +0.77\,V$, so iron(III) ions will react with any halide ions associated with electrode potentials less positive; that is: $I_2(aq) + 2e^- \rightleftharpoons 2I^-(aq)$; $E^\ominus = +0.54\,V$; the only reaction that takes place is $2Fe^{3+}(aq) + 2I^-(aq) \rightarrow 2Fe^{2+}(aq) + I_2(aq)$

(a) (i) No

(ii) Yes

(iii) No

(iv) No

(v) Yes

(b) (i)

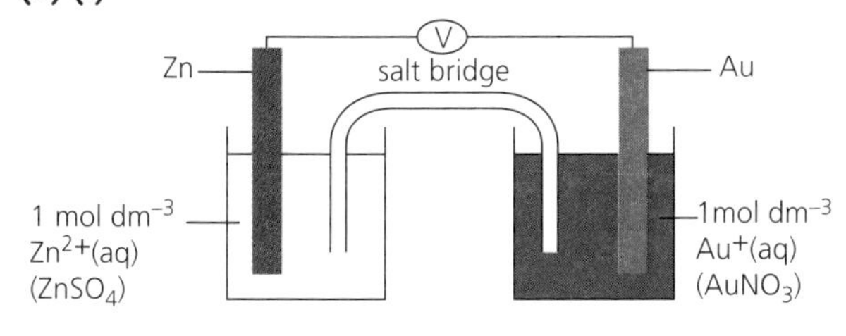

(ii) $E^\ominus_{cell} = +1.68 + (+0.76) = +2.44\,V$

(iii) $Zn(s) \mid Zn^{2+}(aq) \parallel Au^+(aq) \mid Au(s)$

Chapter 13

1 **(a)** 4

(b) 6

(c) 4

(d) 6

(e) 6

2 **(a)** Ethanedioate ions

(b) Octahedral

(c) +3

(d) $[Ni(C_2O_4)_3]^{4-}$

3 **(a)** 2

(b) $[Fe(ind)_3]^{2+}$

4 $Cr_2O_7^{2-} + 14H^+ + 6I^- \rightarrow 3I_2 + 2Cr^{3+} + 7H_2O$

Moles of dichromate(VI) $= \frac{21.25}{1000} \times 0.0100 = 2.125 \times 10^{-4}$ mol

Moles of iodide $= 2.125 \times 10^{-4} \times 6 = 1.275 \times 10^{-3}$ mol

Total moles of iodide $= 1.275 \times 10^{-3} \times \frac{250}{25} = 0.01275$ mol

Mass of iodide $= 0.01275 \times 126.9 = 1.618$ g

So, percentage by mass of iodine as $I^- = \frac{1.618}{50.0} \times 100 = 3.24\%$

5 **(a)** A pale blue precipitate forms;

$[Cu(H_2O)_6]^{2+}(aq) + 2OH^-(aq) \rightarrow [Cu(H_2O)_4(OH)_2](s) + 2H_2O(l)$

or $Cu^{2+}(aq) + 2OH^-(aq) \rightarrow Cu(OH)_2(s)$

This dissolves in excess ammonia to form a dark blue solution;

$[Cu(H_2O)_4(OH)_2](s) + 4NH_3(aq) \rightarrow [Cu(NH_3)_4(H_2O)_2]^{2+}(aq) + 2H_2O(l) + 2OH^-(aq)$

(b) A dark green precipitate forms;

$[Cr(H_2O)_6]^{3+}(aq) + 3OH^-(aq) \rightarrow [Cr(H_2O)_3(OH)_3](s) + 3H_2O(l)$

or $Cr^{3+}(aq) + 3OH^-(aq) \rightarrow Cr(OH)_3(s)$

This dissolves in excess NaOH to form a green solution;

$[Cr(OH)_3(H_2O)_3](s) + 3OH^-(aq) \rightarrow [Cr(OH)_6]^{3-}(aq) + 3H_2O(l)$.

(c) A white precipitate forms and fizzing takes place;

$2[Al(H_2O)_6]^{3+}(aq) + 3CO_3^{2-}(aq) \rightarrow 2[Al(H_2O)_3(OH)_3](s) + 3CO_2(g) + 3H_2O(l)$

6 Sodium carbonate will form a brown precipitate of iron(III) hydroxide with iron(III) and also fizzing takes place; with iron(II) ions, a dark green precipitate of iron(II) carbonate forms and there is no fizzing.

7 **(a)** 6

(b) $[Fe(H_2O)_6]^{2+}(aq) + TPEDA \rightleftharpoons [Fe(TPEDA)]^{2+}(aq) + 6H_2O(l)$

(c) Lots of water molecules are released; so entropy increases in the ligand substitution process; creating a more negative ΔG.